Experiments in
Molecular Biology

Biological Methods

Experiments
in
Molecular Biology

Edited by

Robert J. Slater

Humana Press • Clifton, New Jersey

Library of Congress Cataloging in Publication Data
Main entry under title:

Experiments in molecular biology

(Biological methods)
Includes index.

1. Molecular Biology—Experiments.   I. Slater,
Robert J.   II. Series
QH506.E96   1985   574.87'382   85- 21882
ISBN 0- 89603- 082- 2
© 1986 The Humana Press Inc.
Crescent Manor
PO Box 2148
Clifton, NJ 07015

Printed in the United States of America

PREFACE

Research in the field of molecular biology has progressed at a fascinating rate in recent years. Much of this progress results from the development of new laboratory techniques that allow very precise fractionation and analysis of nucleic acids and proteins, as well as the construction of recombinant DNA molecules that can then be cloned and expressed in host cells. Progress has been so rapid that there has been a shortfall in the training of appropriately qualified staff. Many existing laboratory workers require retraining, and many educational institutions have had difficulty incorporating the new molecular biology techniques into their teaching programs. Although there are several manuals currently available that describe laboratory techniques in molecular biology, they are principally written for the individual research worker and are not intended for use in the design of practical classes for students.

The aim of this book is to provide just such a series of protocols for the teaching of practical molecular biology. The idea arose following the success of several Workshops in Molecular Biology, organized and taught by staff in the Biology Department of the Hatfield Polytechnic. Gradually, the protocols used in the workshops have been incorporated into the Hatfield undergraduate and postgraduate teaching programs and have now been collected together to form a book. The lecturers and demonstrators in these workshops and courses, together with respected authorities from other institutions, have contributed their tried and tested protocols to this book so that others may more readily incorporate material central to modern molecular biology into their own undergraduate and beginning graduate-level courses, or indeed even develop similar laboratory workshops in their own institutions.

The chapters following describe a range of the most important techniques in molecular biology research today. The book begins with chapters on the extraction and manipulation of DNA, continues with articles on RNA and protein techniques, and closes with some experiments on whole cells. In all but the first chapter, which is intended as an introductory experiment, the procedures described are those normally used in the research laboratory and have not been simplified. This was done because the objectives here are to teach participants in the experiments how the techniques are routinely used and to provide a source book of procedures that will be directly useful in their subsequent careers. Nevertheless, the level of detail given is such that every experiment should prove successful at the first attempt.

Each chapter describes an experiment based on a particular technique in molecular biology, and all are presented in the same format. The Introduction gives a brief summary of the techniques to be employed and describes the importance of such techniques to current research. The aims and objectives of the experiment are then described and an indication is given of the time required to complete the practical procedures. The Materials Provided section lists all those items that should be available before the practical work begins. This is not intended to give all the details concerning the equipment or buffers and solutions required, but is a checklist intended to help organization of the experiment and ensure that nothing has been forgotten. The Protocol section describes the practical procedure in detail, in numbered chronological steps. When an experiment involves a number of separate stages, perhaps involving more than one session, the Protocol is divided into discrete sections as appropriate. The Results and Discussion section describes the predicted results and gives details of any data manipulations that are required. In many cases, interesting discussion points are raised, or questions about data interpretation are posed.

This book contains two Appendices. The first of these, Appendix I, supplies general information relevant to many of the preceeding chapters. Appendix II, on the other hand, is divided into sections related to individual chapters and is concerned with the preparatory work required to set up each experiment. Appendix II is self-contained and describes in detail any procedures that must be completed prior to the practical class, lists the special equipment required, and the amounts of materials, solutions, and buffers needed for classes of 20 participants, who will be assumed to work in pairs unless otherwise indicated. Some experiments are best suited to smaller groups, whereas others can be expanded to teach much larger classes; this is indicated where appropriate. It is assumed that basic everyday laboratory equipment, such as pipets, bench-centrifuges, ice buckets, and waterbaths or incubators will be available; these are often not listed.

Each chapter in the book has been written in a style suitable for both teacher and student. If the reader is a teacher of molecular biology, he or she should first select the experiments of interest; the book can then be followed systematically to form a whole course on the techniques of molecular biology, or individual chapters can be chosen as required. The Introduction will give an indication of the time required to carry out the procedures, and Appendix II can be given to the laboratory technician to prepare for the practical class. The participants should first read the Introduction to become familiar with the experiment's purpose. Then, before commencing the practical work, they should read the Materials Provided section to be sure that all the necessary items have been supplied. The experiment may then proceed by following the steps described in the Protocol. Finally, when the course has been completed, we hope and expect that the participants will always feel able to refer to the Protocol section and Appendix II as a resource for future work in the field.

Robert J. Slater

Contents

CHAPTER 3

**Restriction Endonuclease Digestion and
       Agarose Gel Electrophoresis of DNA**

Stephen A. Boffey

CHAPTER 4

Restriction Site Mapping

Alan Hall

CHAPTER 5

Southern Blotting
Alan Hall

CHAPTER 6

Preparative Electrophoresis: *Extraction of Nucleic Acids From Gels*
Alan Hall

CHAPTER 7

DNA Ligation and *Escherichia coli* Transformation

Alan Hall

CHAPTER 8

Detection of Plasmid By Colony Hybridization

Elliot B. Gingold

CHAPTER 9
Microisolation and Microanalysis of Plasmid DNA
Elliot B. Gingold

CHAPTER 10
Isopycnic Centrifugation of DNA: *Extraction and Fractionation of Melon Satellite DNA by Buoyant Density Centrifugation in Cesium Chloride Gradients*
Donald Grierson

CHAPTER 13
Agarose Gel Electrophoresis of RNA
Robert J. Slater

CHAPTER 14
In Vitro Protein Synthesis
Jayne A. Matthews and Raymond A. McKee

CHAPTER 15

**SDS Polyacrylamide Gel Electrophoresis of
    Proteins:** *Determination of Protein Molecular
    Weights and High-Resolution Silver Staining*

John M. Walker

CHAPTER 16

**Isoelectric Focusing (IEF) of Proteins in Thin-
    Layer Polyacrylamide Gels**

John M. Walker

CHAPTER 17

Protein (Western) Blotting With Immunodetection

John M. Walker

CHAPTER 18

Thin-Layer Gel Filtration: *Determination of Protein Molecular Weights*

John M. Walker

CHAPTER 19

The Dansyl-Edman Method for Peptide Sequencing

John M. Walker

CHAPTER 20

The Double Immunodiffusion Technique:
*Immunoprecipitation and Analysis of Antigenic
Protein in Gel*

David L. Cohen

CHAPTER 21

Immunoelectrophoretic Techniques:
*2-D Immunoelectrophoresis and
Rocket Immunoelectrophoresis*

John M. Walker

CHAPTER 22

The Time Course of β-Galactosidase Induction in *Escherichia coli*

Kenneth H. Goulding

CHAPTER 23

Experiments on Enzyme Induction in the Unicellular Green Alga *Chlorella fusca*

Kenneth H. Goulding

CHAPTER 24

Yeast Protoplast Fusion

Elliot B. Gingold

CHAPTER 25

The Ames Test

Elliot B. Gingold

APPENDIX I

Contributors

STEPHEN A. BOFFEY • *Division of Biological and Environmental Studies, The Hatfield Polytechnic, Hertfordshire, UK*

DAVID L. COHEN • *Division of Biological and Environmental Studies, The Hatfield Polytechnic, Hertfordshire, UK*

ELLIOT B. GINGOLD • *Division of Biological and Environmental Studies, The Hatfield Polytechnic, Hertfordshire, UK*

KENNETH H. GOULDING • *Division of Biological and Environmental Studies, The Hatfield Polytechnic, Hertfordshire, UK*

DONALD GRIERSON • *Department of Physiology and Environmental Sciences, University of Nottingham, Loughborough, UK*

ALAN HALL • *Institute of Cancer Research, Chester Beatty Laboratories, London, UK*

RAYMOND A. McKEE • *Department of Botany, University of Leicester, Leicester, UK*

JAYNE A. MATTHEWS • *Department of Clinical Chemistry, Wolfson Research Laboratories, Queen Elizabeth Medical Centre, Birmingham, UK*

ROBERT J. SLATER • *Division of Biological and Environmental Studies, The Hatfield Polytechnic, Hertfordshire, UK*

JOHN M. WALKER • *Division of Biological and Environmental Studies, The Hatfield Polytechnic, Hertfordshire, UK*

Chapter 1

DNA Isolation and Transformation in *Escherichia coli*

An Introduction

Elliot B. Gingold

1. Introduction

The exercise in this chapter is designed for students in the early years of an undergraduate course who have just been introduced to the basic concepts of genetic engineering in theory and would like to supplement their knowledge with practical experience. Many DNA manipulations, however, make use of equipment that is unsuitable for introductory classes with possibly large numbers of students, and require more time than that available in already full timetables. This experiment, therefore, was designed to give students their first feel of handling DNA, though using only the most basic laboratory equipment, and has been carefully organized to fit into a 3-h period. It provides an excellent introduction to the more sophisticated procedures found elsewhere in this volume.

1

The basis of the experiment is that a strain of *Escherichia coli*, carrying the plasmid pBR322 (for structure, *see* Appendix I at end of book) is lysed and a crude DNA extract prepared by precipitating the nucleic acids with cold ethanol, followed by winding the DNA onto a thin glass rod. This DNA is then redissolved in a buffer and used to transform another *E. coli* strain. (Transformation is the aquisition of new genes or alleles following the uptake of nucleic acid into a cell.) The plasmid pBR322 carries genes for ampicillin resistance (Amp^r) and tetracycline resistance (Tc^r). It is the acquisition of these resistances by the originally sensitive recipient strain that is used to detect the transformation.

When plasmid DNA is isolated for research purposes, it is usual to carry out an extensive procedure of purification with the aim of separating the plasmid from all other cellular components. Such an exercise is described in Chapter 2. In this experiment we do not require a high level of purity, since the aim is simply to demonstrate the transforming activity of the plasmid and not to subject it to further biochemical manipulations. Thus the method makes no attempt to separate the plasmid DNA from the chromosomal DNA. Nonetheless, the extract is perfectly adequate for demonstrating transformation.

The procedure used for transformation in this exercise is very closely modeled on normal research methods. Rapidly growing *E. coli* cells are harvested and placed in cold CaCl$_2$. The suspension is then held in ice for 20 min, during which time changes take place in the cell wall of the bacteria, making them "competent" for transformation. The cells are then reharvested and concentrated in CaCl$_2$ in the presence of the DNA sample. At the end of another 20 min on ice, the cells are given a short "heat shock" in a 42°C waterbath. This step has the effect of greatly increasing DNA uptake. After a short incubation in a nonselective medium to allow plasmid establishment and expression, the cells are plated onto antibiotic-containing plates. After incubation, colonies of transformed plasmid-carrying cells should appear on the plates.

With students working in pairs, the experimental procedures should take no more than 3 h. It will be necessary, however, to return to the laboratory the next day to count the colonies growing on the plates, although these can be refrigerated and counted at a later date if that would be more convenient.

2. Materials Provided

2.1. Cultures

> 10 mL Plasmid-carrying strain
> 10 mL Recipient strain

Cultures are supplied in sterile plastic centrifuge tubes.

2.2. Solutions

> TE buffer, pH 8.0
> 20% Sucrose in 5× TE
> 0.5M EDTA, pH 8.5
> 10 mg/mL Lysozyme
> Triton solution
> Absolute ethanol (freezer cooled)
> 0.1M CaCl$_2$

2.3. Growth Media

> LB broth (2 mL in small shake flasks)
> LB agar plate
> LB + ampicillin plates
> LB + tetracycline plates

2.4. Equipment

> Small, sterile glass bottles
> Sterile pipets
> 42°C Waterbath
> 37°C Shaking waterbath
> Ice buckets

3. Protocol

Students will work in pairs, with one member taking responsibility for the DNA isolation and the other for the preparation of the recipient cells.

3.1. DNA Isolation

1. You will be provided with 10 mL of an overnight culture of the plasmid-carrying bacteria in a sterile plastic centrifuge tube. Sediment the cells in a bench centrifuge (5 min at 5000*g*) and discard the supernatant. As with all steps, the supernatant is discarded into flasks that will later be autoclaved.

2. Resuspend the cells in a few drops of cold TE and make up to 10 mL with the same buffer. This stage serves to wash the remaining traces of growth medium from the cells.

3. Centrifuge and collect the cell pellet as before.

4. Resuspend the cells in 0.4 mL of cold 20% sucrose (in 5× TE). Add 0.1 mL of the lysozyme solution.

5. Swirl the tube on ice for 5 min.

6. Add 0.1 mL of cold 0.5*M* EDTA (pH, 8.5). Swirl on ice for a further 5 min. By this stage the cell walls should have been weakened to allow easy lysis.

7. Add 0.8 mL of Triton solution and mix well. This detergent should lyse the cells and produce a viscous suspension.

8. Cool the lysate, then layer 3 mL of freezer-cooled absolute ethanol onto the top. The DNA will precipitate at the interface. RNA and protein precipitates will also form.

9. With a Pasteur pipet, flame-sealed at the tip, gently stir the two layers, stirring in only one direction. Fibers of DNA will wind onto the rod. Continue until no further precipitate is picked up. Much precipitate will remain in the tube, but this is no cause for concern, since this will include most of the RNA and protein.

10. Dry the rod by gently waving in the air in the vicinity of a flame, but not close enough to heat the rod.

11. Redissolve the precipitate by stirring into 0.5 mL of TE in a small glass bottle. This will take a few

minutes. Even with the most careful preparation, some protein will be picked up and this will not redissolve. Thus it is expected that the final solution will be somewhat cloudy.

3.2. Preparing the E. coli Recipient

1. You will by provided with 20 mL of a rapidly growing early log-phase culture of the recipient strain in a sterile plastic centrifuge tube. The cell suspension is cooled in an ice bath for 5 min and then centrifuged for 5 min (5000g).

2. Remove the supernatant and resuspend the cells in 2 mL of cold 0.1M $CaCl_2$. Leave in ice for 20 min.

3. Recentrifuge the suspension as before. The pellet will now appear more diffuse; this is an indication of transformation competent cells.

4. Discard the supernatant and resuspend the cells in 0.4 mL of cold 0.1M $CaCl_2$. The cells can be placed on ice until required.

3.3. Transformation

1. Take two small *glass* bottles and label one $(+)$ and the other $(-)$. (Glass is preferable to plastic, since it allows rapid temperature shifts.)

2. Divide 0.4 mL of the recipient cell suspension between the two bottles.

3. To the $(+)$ bottle add 0.1 mL of the DNA preparation; to the $(-)$ bottle add 0.1 mL of TE.

4. Leave on ice for 20 min.

5. Heat shock the suspensions by transferring to a 42°C waterbath for 2 min.

6. Transfer the contents of each bottle to 2 mL of LB in small conical flasks. Incubate with shaking at 37°C for 30 min. This step is essential to allow the incorporated plasmid to be established and expressed. Ideally, a longer incubation would be preferable.

7. Plate 0.1 mL from each flask onto LB + tetracycline and LB + ampicillin.
8. Plate 0.1 mL of the DNA suspension onto LB
   agar.
9. Incubate the plates at 37°C overnight, and count
   the colonies. Record all results, including plates
   showing no growth.

4. Results and Discussion

1. You should obtain no growth on the (−) plates,
   and a countable number of colonies on the (+)
   plates. The counts on each of the antibiotic (+)
   plates should be similar. Plating the DNA
   preparation should, of course, produce no
   colonies. This plate, however, tests for general
   sterility of your work and also demonstrates that
   no live bacteria survive the DNA isolation procedure.
2. Explain how the results demonstrate transformation, discussing the importance of the controls.
   Are similar counts obtained on each antibiotic
   plate? Is this expected? What would be the
   expected result on an LB + ampicillin + tetracycline plate?
3. Discuss any shortcomings that you feel the
   method has. In particular, does it demonstrate
   that it is actually plasmid DNA that is responsible
   for the transformation? If not, discuss additional
   experiments that could be used to prove this
   point.

Chapter 2

Isolation of Plasmid DNA

The Cleared Lysate and CsCl Gradient Technique

Stephen A. Boffey

1. Introduction

Plasmids first hit the scientific headlines when it was discovered that they are the means by which bacteria are able to transfer properties, such as resistance to an antibiotic, from one cell to another. It is now known that plasmids are circular, double-stranded molecules of non-chromosomal DNA that contain their own origin of replication, and can therefore replicate independently within the bacterial cell. They are considerably smaller than the chromosomal DNA, but can contain genes, not only for antibiotic resistance, but also for antibiotic synthesis, toxin production, nitrogen fixation, production of degradative enzymes, and conjugation. The plasmids can be transferred from one cell to another, and therefore act as carriers or "vectors" of the extrachromosomal genes they contain.

Although plasmids are unwelcome in hospitals, where they can result in the rapid development of multiple resis-

7

tance to antibiotics in a bacterial population, they are of enormous importance in molecular biology. When a plasmid carries its genes into a bacterial cell, it will replicate and be passed on to daughter cells during cell division; thus, many more copies of its genes are produced. If a foreign piece of DNA is inserted into a plasmid and covalently linked to it by ligation, this too will be replicated, along with the rest of the plasmid; thus plasmids can be used as vectors for the cloning of DNA fragments (which may be short sequences, or entire genes).

The essential features of a plasmid, for use in cloning, are as follows. It must contain a replication origin that should if possible, be "relaxed," so that plasmid replication can proceed faster than that of the chromosome. This will result in several molecules of plasmid per cell under normal growth conditions, and further "amplification" of the plasmid can be achieved by the selective inhibition of chromosomal DNA replication. It must be possible to make a single cut in the plasmid, to allow insertion of the foreign DNA, and this can be achieved if there is a unique cleavage site for a particular restriction enzyme. If the plasmid carries a gene for antibiotic resistance, cells containing plasmid can be selected easily; and if the single cleavage site lies within another such gene, then "insertion inactivation" of the gene will allow selection for cells containing plasmid with a piece of inserted DNA. The process of transformation of bacterial cells, in which the cells take up plasmids, is most efficient if the plasmid is small. A small size also aids isolation of the plasmid and minimizes damage by shearing.

Since no naturally occurring plasmid meets all the criteria outlined above, plasmids for use in cloning have been constructed by joining together selected parts of natural plasmids. These plasmids are small, contain a relaxed origin of replication and at least two genes for antibiotic resistance, and have unique cleavage sites for several restriction enzymes, some of which lie within the resistance genes, to allow insertion inactivation. These properties are illustrated by one of the most widely used plasmids, pBR322 (see Appendix I at the end of this book for details of its structure).

The use of plasmids as vectors for the cloning of DNA fragments therefore lies at the heart of genetic manipulation. Before a plasmid can be used as a vector it must, of course, be isolated from its host bacterium. After ligation, transformation, and cloning, the recombinant plasmid must once again be isolated if the inserted DNA is to be recovered.

The cleared lysate method of plasmid isolation is commonly used to extract relatively small plasmids (up to about 20 kb) from gram negative bacteria such as *Escherichia coli*. It relies on a very gentle lysis of the bacteria to release small molecules, including very compact, supercoiled plasmids, into solution, whereas larger molecules, such as chromosomal DNA fragments, are trapped in the remains of the cells. High-speed centrifugation will pellet cell debris and trapped chromosomal DNA to produce a cleared lysate highly enriched for plasmid.

After deproteinization, pure plasmid DNA can be isolated from the lysate using CsCl density-gradient ultracentrifugation in the presence of ethidium bromide. This step is based on the ability of ethidium bromide to decrease the buoyant density of nucleic acids to which it binds, and on the fact that supercoiled plasmid molecules will bind less dye per unit length than linear or relaxed circular molecules. Consequently, supercoiled plasmid will band below all other types of DNA in this ultracentrifugation.

Yields of plasmids such as pBR322, which contain a ColE1 origin of replication, can be increased by amplification using chloramphenicol to inhibit replication of chromosomal DNA, but not of plasmids. Such amplification can result in up to 3000 copies of plasmid per cell. After lysis, clearing the lysate, deproteinization, CsCl density-gradient ultracentrifugation, and dialysis, up to 1 mg of supercoiled plasmid can be obtained from 1 L of bacteria in a form suitable for restriction or transformation. The procedure is developed from that of Clewell and Helinski (1), and requires only 4 h of benchwork, including a 1-h centrifugation and 1 h for ethanol precipitation of nucleic acids; however, owing to a 2-d ultracentrifugation and an overnight dialysis, 3 d are need from start to finish.

The procedure described here is particularly effective when combined with others in this volume demonstrating the techniques of ligation, transformation, agarose gel electrophoresis, and restriction site mapping.

2. Materials Provided

2.1. Solutions and Chemicals

Tris/sucrose
Lysozyme, 5 mg/mL in 0.25M Tris, pH 8.0
EDTA 0.25M, pH 8.0
Sodium dodecyl sulfate (SDS), 10% (w/v)
Tris 0.25M, pH 8.0
Tris-base 2.0M, no pH adjustment
Phenol/chloroform mixture, 1:1 (v/v)
Potassium acetate 4.4M
Absolute ethanol, stored at $-20°C$
70% Ethanol
0.1× SSC
CsCl
Ethidium bromide 5 mg/mL
Isoamyl alcohol

2.2. Equipment

Ice bucket
Vortex mixer
Conical flask, 100 mL
Waterbath, 37°C
Thick-walled polycarbonate centrifuge tubes, 25 or 35 mL
Measuring cylinder, 10 mL
Thick-walled glass centrifuge tubes, 15 mL
Plugged Pasteur pipets
Polypropylene centrifuge tube, 50 mL, plus cap
High-speed centrifuge and angle rotor for at least six 50-mL tubes
Vacuum desiccator, or nitrogen gas cylinder
Disposable gloves and safety glasses
Ultracentrifuge tube, 5 mL, UV transparent
Ultracentrifuge and swing-out rotor to take six tubes
UV lamp (long wavelength)
Syringe, 1 mL, and wide-bore needle
1.5-mL microfuge tube
Dialysis tubing, pretreated with EDTA

3. Protocol

1. Decant the supernatant from a tube of centrifuged cells into a disposal bin (remember you are work-

ing with living cells at this stage). Seal the tube with a cap and vortex mix the pellet until it forms a smooth paste.

2. Add 3.75 mL of Tris/sucrose to the paste, and mix to give a homogeneous suspension. Transfer this suspension into a precooled 100-mL conical flask.

3. Add 0.75 mL of lysozyme solution, then 1.25 mL of EDTA (0.25M, pH 8). Swirl on ice for 10 min to allow the EDTA and lysozyme to weaken cell walls.

4. Add 1.5 mL of 10% SDS (giving a 2% solution), and immediately give the flask a single swirl to ensure mixing. After this, it is important to treat the suspension as gently as possible to avoid damaging high molecular weight DNA released from the cells.

5. Incubate at 37°C, without shaking, until the suspension loses its cloudy appearance as a result of cell lysis. This usually occurs within a few minutes and may be confirmed by holding the flask at eye level, tilting it gently, and watching for a highly viscous "tail" sliding down the glass behind the bulk of the suspension. This high viscosity results from the release of high molecular weight DNA from the bacteria.

6. Tip this lysate gently into a 25- or 35-mL, thick-walled polycarbonate centrifuge tube. Balance tubes in pairs, using Tris (0.25M, pH 8.0) if necessary, and centrifuge them at 120,000g for 1 h at 20°C. This will clear the lysate by forming a translucent pellet containing cell debris tangled with much of the high molecular weight chromosomal DNA. Consequently, the supernatant will contain most of the plasmid and relatively little chromosomal material; it will also contain RNA and proteins.

7. Carefully decant each supernatant into a measuring cylinder, note its volume, and transfer into a 100-mL conical flask.

8. To the supernatant add 0.1 times its volume of 2.0M Tris-base (pH not adjusted). Then double the volume by adding phenol:chloroform (1:1, v/v).

Shake this mixture sufficiently to form an emulsion, and keep it emulsified by occasional shaking for 10 min.

Note: *Be very careful when handling phenol. Wear rubber gloves and safety glasses, and mop up spills using ethanol; wash well with water if skin is splashed.*

9. Centrifuge the emulsion in thick-walled glass centrifuge tubes at top speed in a bench centrifuge (about 3500g) for 10 min to separate the aqueous and organic phases. Transfer the upper (aqueous) layer into a clean measuring cylinder using a short Pasteur pipet; be careful not to transfer any of the white precipitate at the interface (this contains denatured protein).

10. Add 0.25 times the volume of 4.5M potassium acetate (to give a 0.9M solution). This is needed to ensure quantitative precipitation of low concentrations of DNA in the next step.

11. Calculate the new volume of solution, and transfer it to a 50-mL polypropylene centrifuge tube. Add 2 vol of chilled ethanol and mix thoroughly; then leave the capped tube at $-20°C$ for at least 1 h to allow precipitation of DNA and some RNA.

12. Centrifuge at 12,000g for 10 min at 0°C, using 70% ethanol to adjust tubes to balance in pairs. Decant the supernatant and, keeping the tube inverted, blot off any remaining drops of ethanol. Remove traces of ethanol by evaporation in a vacuum desiccator, or by gently blowing a stream of nitrogen into the tube through a sterile Pasteur pipet plugged with cotton wool.

13. Dissolve the precipitate, which may be invisible, in 3.6 mL of 0.1 × SSC by gently swirling the liquid round the tube. Do not use a vortex mixer for this step. When combining this experiment with that on agarose gel electrophoresis, it is convenient to dissolve the precipitate initially in 0.4 mL of 0.1 × SSC and to take a 20 μL sample for subsequent electrophoresis; then make up the remaining solution to 3.6 mL.

14. Transfer the solution into a flask containing 3.9 g CsCl, and mix until this is completely dissolved. Then add 0.4 mL of ethidium bromide (5 mg/mL). *Wear gloves for this, since ethidium bromide may be carcinogenic.*

15. The solution is now centrifuged for at least 40 h at 140,000g in a swing-out rotor, at 20°C. Rotors that take short, wide tubes give the sharpest bands.

16. After centrifugation, view the tubes using long-wavelength UV light. Two well-defined bands should be seen near the middle of the tube, separated by about 1 cm. The lower band consists of supercoiled, covalently closed, circular plasmid. If the cleared lysate procedure has worked well, this band should be more intense than the upper one, which contains fragments of chromosomal DNA and also linear and open circle forms of plasmid.

17. To recover the plasmid band, first draw off the upper part of the gradient, including the upper DNA band, using a Pasteur pipet. Then fix the centrifuge tube beneath a syringe fitted with a wide-bore needle and lower the syringe (or raise the tube) slowly until the needle tip is just below the plasmid band. Provided both syringe and tube are firmly fixed, you should have no difficultly in drawing the plasmid band into the syringe. When all the plasmid has been removed, a very thin band will remain in the tube; this is an optical effect caused by the sudden change in refractive index where a slice of continuous gradient has been removed. Do not try to collect this "band"!

18. Transfer the plasmid material into a 1.5-mL polypropylene tube (approximately 0.5 mL/tube), and add almost enough isoamyl alcohol to fill each tube. Cap the tubes and invert them several times. The alcohol will become pink as ethidium bromide partitions into it. Remove the upper (alcohol) layer, and re-extract with fresh alcohol.

Repeat this until no color can be detected in the alcohol, then once more to be sure.

19. To remove CsCl from the plasmid solution, transfer it into dialysis tubing (pretreated by boiling in 10 mM EDTA for 15 min, followed by two 15-min treatments in boiling distilled water) and dialyze against three changes of 500 mL 0.1 × SSC over about 16 h (i.e., overnight) at 4°C. Wear disposable gloves when handling the dialysis tubing.

20. The plasmid is now ready for use. If a more concentrated preparation is needed, concentrate it by precipitation with ethanol, as described in steps 10–12. In its present form, it can be used for transformation, as described in Chapters 1 or 7, or it can be examined by agarose gel electrophoresis both before and after digestion with restriction endonucleases (Chapter 3).

4. Results and Discussion

Although this procedure does not include an assay of the plasmid, there are several stages at which it is possible to tell if the preparation is going according to plan. After treatment with SDS, the bacterial suspension should no longer be cloudy, but should become highly viscous and even gel-like. This indicates that the cells have been successfully lysed. More clues are provided by the appearance of the pellet after clearing the lysate. Excessive lysis and shearing of DNA will result in a very small pellet in which little chromosomal DNA is trapped; consequently the solution will not have been significantly enriched for plasmid.

After CsCl ultracentrifugation, it is easy to tell if the procedure has been successful. Two DNA bands should be clearly visible in UV light, and the lower, supercoiled plasmid band should be the more intense in a good preparation.

Analysis of the extracted, dialysed product by agarose gel electrophoresis (Chapter 3) is the quickest way to test the quality of the plasmid.

Reference

1. Clewell, D.B. and Helinski, D.R. (1971) Properties of a super-coiled deoxyribonucleic acid-protein relaxation complex and strand specificity of the relaxation event. *Biochemistry* **9**, 4428–4440.

Chapter 3

Restriction Endonuclease Digestion and Agarose Gel Electrophoresis of DNA

Stephen A. Boffey

1. Introduction

It is no exaggeration to say that genetic engineering has been made possible by the discovery and isolation of enzymes that recognize specific nucleotide sequences in DNA and can cleave the DNA at a precise point within the sequence. These enzymes are known as Type II restriction endonucleases, and they allow us to cut DNA molecules in a reproducible fashion. This precision cutting is essential for many of the procedures of genetic manipulation. Circular plasmid molecules must be cut at a unique site, to open them prior to the insertion of foreign DNA for cloning. For the isolation of genes, all molecules of the genomic DNA must be cut identically, so that each gives the same set of restriction fragments. With luck, one of these fragments will contain the desired gene, and further restriction of the isolated fragment may allow recovery of a smaller fragment containing little more than the gene. After cloning of such DNA fragments, they are usually excised from the vector

17

DNA, and this again uses restriction enzymes. There are many more ways in which these enzymes are used by molecular biologists (Chapter 4 describes their use for mapping DNA molecules), but their importance should already be apparent. For further information on restriction endonucleases, *see* Old and Primrose (1).

Another technique that is central to genetic manipulation is the electrophoresis of DNA in agarose gels. This process separates DNA molecules from each other on the basis of their size, and the molecules are made visible using the fluorescent dye ethidium bromide. In this way, DNA can be checked for size, intactness, homogeneity, and purity. Restriction fragments can be separated from each other, prior to identification by such procedures as Southern blotting and hybridization, and the separated fragments can be recovered from the gel (*see* Chapters 5 and 6) for cloning, sequencing, and other procedures.

Agarose forms a gel by hydrogen bonding when in a cool aqueous solution, and the gel pore size depends on the agarose concentration. When DNA molecules are moved through such a gel under the influence of a steady electric field, their speed of movement depends almost entirely on their size, the smallest molecules having the highest mobilities. This relationship holds over a limited range of sizes, and so gel concentration must be chosen to suit the size range of the molecules to be separated. For DNA between 0.5 and 10 kilobase pairs (kb), 0.8% gels are suitable. Such gels can be used to measure the sizes of linear DNA molecules. A calibration curve is produced by plotting mobilities against the log molecular weight of suitable markers, and this is then used to determine the lengths of unknown DNA molecules whose mobilities lie within the linear part of the calibration curve. Such calibrations cannot be extended from linear DNA to circular forms, because linear, open-circle, and supercoiled forms of the same DNA will have markedly different mobilities.

A variety of systems has been described for the electrophoresis of DNA, but possibly the simplest is the "submarine" or "submerged" gel, as described by Maniatis et al. (2). This is based on a horizontal gel that is completely submerged in buffer during electrophoresis. In spite of its simplicity, this system gives excellent results, is easy to use, and can be constructed easily and cheaply; it is therefore very popular. (An alternative system using wicks is described in Chapter 5.)

The experiment described here combines restriction analysis with agarose gel electrophoresis, using restriction endonucleases to produce DNA fragments of known size from λ DNA, and making use of these fragments as markers for measuring the size of plasmid pBR322. The experiment also demonstrates that different physical forms of the same DNA can have different mobilities on electrophoresis in agarose. A set of incubation mixtures is prepared containing λ DNA or pBR322, with or without restriction enzyme, and these are left for 1 h to allow restriction to proceed to completion. During this time, the agarose gel is poured and allowed to set. The incubated samples are then loaded onto the gel and electrophoresed overnight. The next morning, the gel is stained for about 30 min and then photographed. Measurements of mobilities can be made from the photograph whenever convenient.

There are many ways in which this experiment could be varied. It is interesting to combine it with a plasmid isolation (Chapter 2) and run a sample of the crude plasmid (after ethanol precipitation, but before CsCl ultracentrifugation) in one of the unused gel tracks. This will show how successful the cleared lysate method is in enriching for plasmid, and will reveal that the sample is highly contaminated with RNA. It will also show how careful the students have been in handling the DNA, as indicated by the proportions of supercoiled, relaxed, and linear plasmid.

If the equipment is available, electrophoresis may also be performed in vertical gels, using the same buffer and gel solutions, and the resolution and ease of use of the two types of apparatus can be compared. Vertical gel apparatus for agarose gels resembles that used for polyacrylamide gels (see Chapter 15), but uses thicker spacers and combs and may rely on frosted glass plates or a net at the bottom of the plates to keep the gel in place. Very rapid separations may be achieved using a "minigel" system, such as the BioRad Mini-Sub-DNA Cell. In this way, the time needed for electrophoresis can be reduced to 2 h, and the whole experiment can be carried out in 1 d. The use of minigels is described in Chapter 8.

2. Materials Provided

2.1. Solutions and Chemicals

Alcohol in wash bottles
Gel-running buffer, diluted to working concentration
Agarose solution, 100 mL portions, kept at 50°C
DNA from bacteriophage λ, on ice
Plasmid pBR322, on ice
Restriction endonuclease EcoRI, on ice
Restriction endonuclease HindIII, on ice
Restriction buffer, on ice
Sterile distilled water
Gel-loading solution
Ethidium bromide

2.2. Equipment

Glass casting plate
Zinc oxide tape
Scissors
Leveling table and spirit level (optional)
Well-forming comb
Microcentrifuge tubes
Rack for microcentrifuge tubes
Adjustable, automatic micropipet
Disposable tips for micropipet
Microcentrifuge, with rotor to take 0.5-mL tubes
Water bath, 37°C
Electrophoresis tank
Power supply for electrophoresis
Shallow tray or plastic box, for staining gel
Pair of disposable gloves
Transilluminator
Safety glasses
Camera, filters, camera stand, film

3. Protocol

3.1. Restriction of DNA

1. Place six numbered microcentrifuge tubes in a rack, and pipet the solutions indicated in Table 1

Table 1
Contents of Microcentrifuge Tubes

| | Tube number | | | | | |
Sample	1	2	3	4	5	6
λ DNA	2 μL	5 μL	5 μL	5 μL	—	—
pBR322	—	—	—	—	5 μL	5 μL
Restriction buffer	2 μL	2 μL	2 μL	2 μL	2 μL	2 μL
Sterile water	16 μL	11 μL	11 μL	9 μL	13 μL	11 μL
EcoRI	—	2 μL	—	2 μL	—	2 μL
HindIII	—	—	2 μL	2 μL	—	—

into each, using an automatic micropipet fitted with a sterile tip. Use a fresh tip for each solution, and be careful not to transfer solution from one microcentrifuge tube to another.

2. When all additions have been made, close each tube and mix its contents using a few sharp taps against the side of the tube. Spin the tubes for about 1 s in a microcentrifuge to drive their contents to the tube tips. Repeat the mixing and centrifugation once.

3. Place the rack of tubes in a 37°C water bath for 1 h to allow digestion to proceed to completion.

4. After the incubation, add at least 2 μL of loading solution to each tube. It is almost impossible to pipet this solution accurately, owing to its very high viscosity; if in doubt, be generous. Mix the solutions thoroughly, as in step 2 above, being particularly careful to ensure that the dense loading solution mixes completely with the incubated solutions.

3.2. Casting of Agarose Gel

1. Wash the glass casting plate carefully, using alcohol to remove any grease. Dry the plate thoroughly, then form a wall around it using zinc oxide tape. Press the tape firmly against the edge of the glass to ensure firm attachment, paying particular attention to the corners of the plate and the region where the tape ends overlap; there

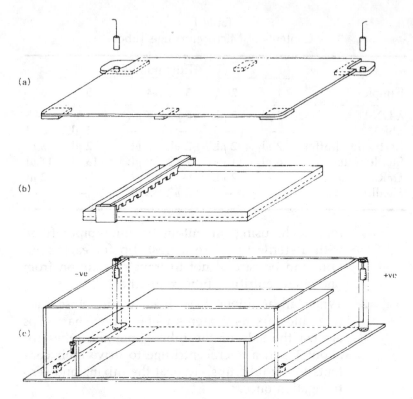

Fig. 1. Apparatus for submerged agarose gel electrophoresis.
(a) Lid, with holes for electrical connectors; (b) Casting plate bor-
dered by zinc oxide tape. The well-forming comb is shown in
position, and should be removed before placing the gel, still on its
plate, in the tank; (c) Electrophoresis tank, with raised platform
for gel, and large volume electrode chambers at each end.

The apparatus is most easily made from perspex, although
glass must be used for the casting plate, and platinum wire for
the electrodes. Dimensions can be varied to suit particular
requirements, and will be determined by the size of casting plate
to be used. Connectors for the power supply should be attached
to the lid so that the apparatus cannot be run with its lid
removed.

> should be no tape folded underneath the plate.
> (Do not stretch the tape, or it will bow in along
> the sides of the plate.) This should result in a
> leakproof wall about 9 mm high all around the
> plate, as shown in Fig. 1.

2. Place the prepared plate on a bench or leveling
   table, and check with a spirit level to see that it is
   perfectly horizontal.

3. Take 100 mL of agarose solution from the 50°C
    water bath and pour it onto the casting plate; do
    not allow the solution to cool before pouring it
    out or it may set unevenly. As soon as the
    agarose has been poured, place the well-forming
    comb in position, about 3 cm from one end of the
    plate, ensuring the comb is exactly at right angles
    to the sides of the plate and is free of air bubbles.
    Leave at room temperature until the gel has set;
    this takes about 0.5 h, and the gel will then
    appear cloudy.

3.3. *Loading and Running Gel*

1. Gently remove the comb from the set gel and
    transfer the gel, still on its glass plate, into the
    electrophoresis tank, placing its sample wells near
    to the cathode (negative electrode). Pour running
    buffer into the tank until the gel is submerged
    and the buffer is about 1 mm above the zinc
    oxide tape. Note that the tape has been left in
    place to prevent unwanted movement of the gel;
    it has no effect on the electrophoresis.

2. Samples are loaded into the wells using a micro-
    pipet or microsyringe. With the syringe or pipet,
    tip a couple of millimeters above the well, gently
    dispense the sample, which should fall into the
    well owing to its high density. This needs rea-
    sonably steady hands, but it should be quite easy
    to load the whole of each sample into a single
    well.

3. When all samples are loaded, the apparatus is
    closed, connected to a power pack (check that the
    negative output is connected to the electrode
    nearer the sample wells), and run at 40 V over-
    night.

4. After the gel has run for about 16 h, turn off and
    disconnect the power supply. Transfer the gel on
    its glass plate into a shallow tray or plastic box.
    Remove the zinc oxide tape, cover the gel with
    250 mL of used gel-running buffer, and then add
    50 μL of ethidium bromide (5 mg/mL) to the

solution around the gel (not on top of the gel). Mix well, by rocking the tray. *Wear disposable gloves when handling this solution or stained gels; ethidium bromide is mutagenic.*

5. Allow the gel to stain for about half an hour, then transfer it onto a UV transilluminator. Since glass blocks UV light, the gel must be placed either directly on the transilluminator filter or on a sheet of plastic wrap. Place a perspex screen over the gel, put on safety glasses, and turn on the UV light. Nucleic acids on the gel will appear orange, owing to the fluorescence of bound ethidium bromide. In order to prolong the life of the transilluminator, and to avoid excessive exposure to UV radiation, it is best to photograph the gel as soon as it has been checked for the presence of bands, and to use the photograph for detailed interpretation of the restriction patterns.

4. Results and Discussion

After staining, the appearance of the gel should be similar to that in Fig. 2. By reference to the lists of restriction fragment sizes given in Table 2, it should be possible to identify each band produced by restriction of the λ DNA. It is rarely possible to detect the smallest bands using the recommended quantities of DNA; however, more DNA would result in overloading of the larger bands, which are more important for the purpose of this experiment. Since the two ends of λ DNA are "cohesive," the restriction fragments produced from each end tend to stick together at room temperature. Using either EcoRI or HindIII, the largest fragment is derived from one end of λ DNA, and one of the smallest fragments comes from the other end. Consequently an extra band may be resolved, very slightly larger than the largest "authentic" band, formed by the association of the two "end" fragments. The resulting doublet can be seen in tracks 2–4 of Fig. 2.

Measure the distance moved by each band from the front of the loading well, and plot these distances against log molecular weight, to give a calibration curve. Note that

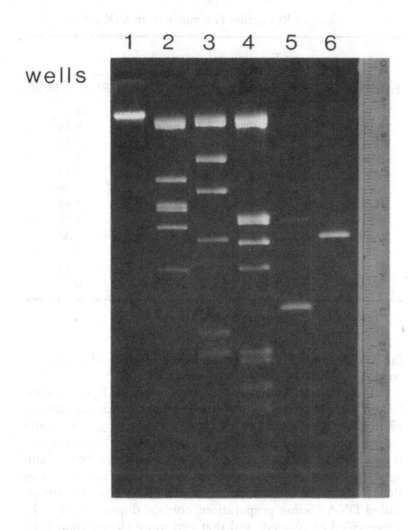

Fig. 2. Results of agarose gel electrophoresis. Samples were prepared as described in the text. After electrophoresis for 16 h at 40 V, the gel was stained with ethidium bromide and photographed using UV illumination.

Contents of tracks: 1, undigested λ; 2, λ digested with EcoRI; 3, λ digested with HindIII; 4, λ digested with both EcoRI and HindIII; 5, undigested pBR322, showing mainly supercoiled plasmid, but also a little open circular form; 6, pBR322 digested with EcoRI.

The scale to the right of the gel is marked in centimeters, with millimeter subdivisions (samples prepared by Matthew C. Jones, Hatfield).

Table 2
Sizes of Restriction Fragments from λ DNA[a]

Size, kilobase pairs (kb)		
EcoRI	HindIII	EcoRI plus HindIII
21.8	23.7	21.8
7.52	9.46	5.24
5.93	6.75	5.05
5.54	4.26	4.21
4.80	2.26	3.41
3.41	1.98	1.98
		1.90
		1.71
		1.32
		0.93
		0.84
		0.58

[a] Intact λ DNA has a size of 49 kb.

the line produced is not linear at the extremes of molecular weight.

Depending on the quality of the pBR322, the unrestricted sample may produce one or more bands on electrophoresis. Ideally, a single band containing covalently closed, supercoiled plasmid should be seen. In practice, it is likely that some of the supercoils will have been damaged, resulting in the formation of open (relaxed) circles that have a lower mobility than the very compact supercoiled DNA. Some preparations contain dimers that can be supercoiled or relaxed, and that run more slowly than their corresponding monomers.

Even if many forms of plasmid are present in the unrestricted sample, only one band should be seen after restriction with EcoRI or HindIII, since each of these enzymes has a unique cleavage site on the plasmid. Consequently, all forms of the plasmid are converted to the monomeric, linear form. The mobility of this linear DNA should be measured and its size read from the calibration curve. It may be possible to see a trace of linear plasmid in the unrestricted sample, if this has been badly damaged.

References

1. Old, R.W. and Primrose, S.B. (1982) *Principles of Gene Manipulation: An Introduction to Genetic Engineering.* 2nd Ed. Blackwell, Oxford.
2. Maniatis, T., Fritsch, E.F., and Sambrook, J. (1982) *Molecular Cloning: A Laboratory Manual.* Cold Spring Harbor Laboratory, New York.

Chapter 4

Restriction Site Mapping

Alan Hall

1. Introduction

The first preliminary characterization of any newly isolated DNA fragment usually involves restriction site mapping. For this, a family of bacterial enzymes called restriction endonucleases, or restriction enzymes, are utilized. These enzymes recognize specific sequences of DNA, usually four or six base pairs, and cut the DNA at that sequence. For example, the restriction enzyme *Eco*RI recognizes the sequence:

$$\downarrow$$
5'-GAATTC-3'
3'-CTTAAG-5'
$$\uparrow$$

and makes a staggered cut in both strands of the DNA at the positions shown by the arrows. Every time that particular sequence occurs in a fragment of DNA, *Eco*RI will make a cut and generate a number of smaller fragments. Restriction site mapping entails locating the positions along a stretch of DNA that are recognized and cleaved by a set of restriction enzymes. Almost 300 restriction enzymes covering 100 or so different recognition sequences are now known. However, only a subset of about three to six

29

enzymes are usually used for mapping in the first instance. This is a first-order analysis of the DNA sequence (the next step being the total sequencing of the DNA), and serves as a fingerprint for the particular DNA fragment. This fingerprint can then be used in a variety of ways, e.g., to compare with other published DNAs, as a preliminary analysis for DNA sequencing, or to localize genes or other structural features to specific regions of a cloned DNA fragment.

Several methods have been developed to facilitate and speed up detailed restriction mapping. In the first instance, restriction enzymes with hexanucleotide recognition sequences, "six-base cutters," are used because they cut the DNA relatively infrequently, and consequently give simple maps. The easiest mapping technique involves digesting the DNA with a series of six-base cutters individually and then in various combinations. The digested DNAs are electrophoresed through an agarose gel along with known DNA size markers. The gel is then stained with ethidium bromide and the DNA is visualized under UV light. After calculating the sizes of the fragments obtained from digestion, it is generally a straightforward matter to fit the data unambiguously into a unique restriction map. As an illustration, enzyme A cuts a fragment of DNA to yield two new fragments with lengths 0.8 and 0.2 units (*see* Fig. 1a). Enzyme B also cuts the same DNA to give two fragments, but with sizes 0.1 and 0.9 units (Fig. 1b). There are, therefore, two possible restriction maps for enzymes A and B, namely c and d. The two are easily resolved by carrying out a double digest using both enzymes. It can be seen that lengths of 0.1, 0.1, and 0.8 units will be obtained in case c, and 0.1, 0.7, and 0.2 units in case d. In this way relatively complicated maps of six or more different enzymes can be built up.

If an enzyme cuts many times, analysis of the data becomes more difficult. Figure 1e–j shows the six possible maps for an enzyme A that cuts a nonsymmetrical fragment in two places to yield fragments with sizes of 0.1, 0.3, and 0.6 units (the more it cuts the more complicated it becomes). One way of simplifying this problem is to partially digest the DNA with enzyme A. New fragments will appear in the agarose gel in which one or other of the sites for A has been left intact. Thus, in addition to the 1.0 (undigested), and the 0.1, 0.3, and 0.6 (totally digested)

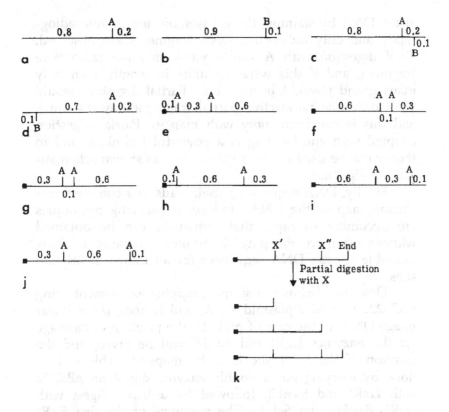

Fig. 1. Schematic representation of restriction maps: a, b, c, and d, two enzymes (A and B), each cleave the fragment once; e, f, g, h, i, and j, one enzyme (A), cleaves a fragment twice; k shows the three cleavage products observed after electrophoresis and autoradiography of a fragment that has been radioactively labeled at one end (the black box) and subjected to partial digestion with enzyme X. It can be seen that the smallest *radioactively labeled* fragment observed corresponds to enzyme X cutting at the nearest site to the end.

bands, partial digestion might show bands of 0.7 and 0.4. Maps e, h, i, and j can then be ruled out, and the map must be either g or f.

Another technique makes use of end-labeling DNA with radioactive nucleotides. With this, albeit slightly more involved technique, the correct map from Fig. 1e–j could be obtained unambiguously. For example, the end marked with a black square is labeled with ^{32}P, and the radioactive DNA is again subjected to either total or partial digestion with enzyme A. After digestion, the DNA is electrophoresed in an agarose gel, but instead of looking directly at the DNA by staining, the gel is subjected to autoradiography and only radioactive DNA fragments are observed. Total digestion with A would yield only *one* radioactive fragment, and if this were 0.3 units in length, then only maps g and j would be possible. Partial digestion would yield only one radioactive partial fragment, say 0.4 units, and this is consistent only with map g. Partial digestion coupled with end-labeling is a powerful technique and in theory can be used to order many sites, as shown schematically in Fig. 1k.

Finally, DNA sequencing itself leads to a complete restriction map of the DNA. Indeed, sequencing techniques are becoming so rapid that sequences can be obtained without extensive mapping. Computer programs are now available to scan DNA sequences for all known restriction sites.

Described below is a simple mapping experiment using pBR322, a circular plasmid DNA, and lambda (λ), a linear phage DNA. In the case of pBR322, the position of cleavage for the enzymes *Eco*RI and *Bam*Hl will be given, and the position of the *Sal* 1 site is to be mapped. This will be done by carrying out a double enzyme digest on pBR322 with *Eco*RI and *Bam*Hl, followed by a triple digest with *Eco*RI, *Bam*Hl, and *Sal* 1. The positions of the five *Eco*RI sites on λ DNA are also presented, and again the *Sal* 1 site is to be mapped. (There are actually two *Sal* 1 sites in λ, but they are so close together that they behave as one site in this experiment.) This will be done using single-enzyme digests of λ DNA with *Eco*RI or *Sal* 1, followed by a double digest with *Eco*RI and *Sal* 1. The experiment will take 1.5 d to perform, the digestions being done on the first day and the gel loaded and run overnight. The following morning the gel will be stained and photographed and the results may then be calculated.

2. Materials Provided

2.1. Solutions

Distilled and deionized water for digestions
10× Concentrated digestion buffer
pBR322 DNA (0.1 mg/mL)

λ DNA (0.1 mg/mL)
*Eco*RI Enzyme (1 unit/µL)
*Bam*Hl Enzyme (1 unit/µL)
Sal 1 Enzyme (1 unit/µL)
Agarose gel-loading buffer

2.2. Agarose Gel Electrophoresis

Agarose powder or solution
10× Concentrated electrophoresis buffer
Deionized water
1 µg/mL Ethidium bromide

2.3. Apparatus and Equipment

11 Microfuge tubes (1.5 mL)
Tube rack
Glass plate and gel slot former
Tape
Bunsen burner and stand for boiling agarose
Power pack
Electrophoresis buffer tanks
Large tray for staining the gel
Polaroid camera and film
UV light box
Plastic face mask
Waterbaths, 37°C and 70°C
Semilog graph paper

3.   Protocol

Enzyme digestions are carried out in 1.5-mL microfuge tubes. These are convenient since many tubes can be centrifuged together in a small benchtop centrifuge. Solutions are aliquoted using automatic pipet tips and a variable automatic pipet. Tips are not generally used more than once, to prevent cross-contamination.

3.1. Day 1

1. Prepare the gel mould for an agarose gel as described in Chapter 3, section 3.2. Add the following to a 500-mL conical flask:

1 g Agarose powder

10 mL 10× Concentrated electrophoresis buffer

90 mL Deionized water

Boil for 2 min or autoclave to melt the agarose. Allow the agarose solution to cool to hand-hot, then pour into the gel mould and place the comb as shown in Chapter 3, Fig. 1. This can be setting while the reactions are being carried out.

2. Label 11 tubes 1–11 and place in a suitable rack. Pipet the correct amount (Table 1) of deionized, distilled water into each tube. (Here the same tip may be used each time since there is no chance of cross-contamination.) Add the 10× concentrated digestion buffer and the DNA. Use new tips each time.

Table 1
Procedures for Restriction Site Mapping Experiment

Solutions	pBR322 Samples						λ Samples				
	1	2	3	4	5	6	7	8	9	10	11
H_2O (μL)[a]	31	30	30	30	29	28	31	30	30	29	22
10 × concentrated digestion buffer (μL)[a]	4	4	4	4	4	4	4	4	4	4	4
pBR322 DNA (μL) (0.1 mg/mL)	5	5	5	5	5	5	—	—	—	—	—
λ DNA (0.1 mg/mL)	—	—	—	—	—	—	5	5	5	5	10
EcoRI (1 unit/μL) (μL)	—	1	—	—	1	1	—	1	—	1	2
BamHl (1 unit/μL) (μL)	—	—	1	—	—	1	—	—	—	—	2
Sal 1 (1 unit/μL) (μL)	—	—	—	1	1	1	—	—	1	1	—
Total volume (μL)[a]	40	40	40	40	40	40	40	40	40	40	40

[a] Precise volumes may be varied according to the size of the wells in the agarose gel.

3. Close the cap and mix the contents. This is usually done by gently flicking the bottom of the tube with one finger while holding the tube in the other hand. Spin for 5 s in the small benchtop centrifuge to bring the contents to the bottom of the tube.

4. Add the relevant enzyme to each tube (Table 1) and again close the cap. The enzyme will be stored in glycerol and will consequently fall to the bottom of the tube when added. It is essential to mix well. However, do not vortex since this can denature proteins. One unit of enzyme, which is defined as the amount of enzyme required to digest 1 μg of DNA in 1 h, is added. Since only 0.5 μg of DNA is used in these experiments, this amount should be sufficient to give complete digestion. Again centrifuge for 5 s to bring the contents to the bottom.

5. Incubate for 1 h in a 37°C waterbath.

6. Stop the reaction by adding 10 μL of agarose gel-loading buffer and mixing well. This contains EDTA, which will chelate the Mg^{2+}—an essential cofactor for the restriction enzymes. The buffer also contains sucrose and the dye bromophenol blue to facilitate loading the samples on a gel.

7. Heat samples 7–11 at 70°C for 10 min, then put on ice for 5 min. Set up the gel for electrophoresis as described in Chapter 3, load all the samples onto the 1% agarose gel in the correct order (1–11), and electrophoresis at 40 V overnight.

3.2. Day 2

1. Place the gel, still on the glass plate, in a container with ethidium bromide so that the gel is completely covered. Agitate gently for 30 min to stain the DNA. *Always wear gloves when handling ethidium bromide.* Replace the ethidium bromide with water and agitate again for 10 min to destain the gel. Slide the gel off the glass plate onto a piece of plastic wrap and observe the DNA under

a UV lamp or with a transilluminator. *Ultraviolet radiation can damage eyes and skin, so always wear a plastic face mask.* Photograph the gel.

2. Measure the distances moved for all fragments in tracks 2–6 and 8–11.

4. Results and Discussion

Track 11 is an *Eco*RI + *Bam*HI digest of λ, and the known sizes of the fragments generated are given in Fig. 2a. By plotting the log molecular weight against distance of migration, a straight line should be obtained. Semi-log graph paper is used. It will become apparent that very large fragments (>15 kb) and very small fragments (<800 bp) will deviate from the straight line. In order to measure such sizes more accurately, a lower percentage gel can be used for the large fragments, and a higher percentage gel, or even better an acrylamide gel, for the smaller fragments. However, the 1% gel will suffice here.

Track 1 contains undigested pBR322. This will be mainly in the form of monomeric supercoils (so called Form I, *see* Fig. 2b), and is the fastest migrating molecule present in this lane. Some nicked open-circular DNA will also be present (Form II), and this will migrate slower. Linear pBR322 (Form III) (which runs a little faster than Form II but much slower than Form I) should be seen in tracks 2, 3, and 4 since all three enzymes used have only one site for cutting.

For circular DNA molecules like pBR322, which have only one *Eco*RI site, this site is usually taken as a point of reference. *Eco*RI recognizes the sequence:

5′-GAATTC-3′
3′-CTTAAG-5′

and the first T is taken as base pair number 1 (Fig. 2c). There are 4362 base pairs in pBR322 (*1*). Therefore the single band obtained in track 2 should measure approximately 4.3 kb when the distance traveled by this fragment is fitted on the marker graph plot. Similarly, there is only one site

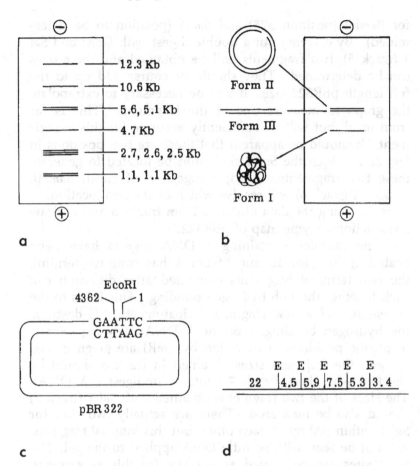

Fig. 2. a, Sizes of fragments (in kilobase pairs, kb) generated after double digestion of λ DNA with EcoRI and BamHI. Some of the fragments are so close in size as to be indistinguishable on an agarose gel; b, Relative electrophoretic mobilities of the three different forms of plasmid DNA. The fastest-moving molecule is supercoiled plasmid (Form I). Nicked open circular DNA (Form II) and linear DNA (Form III) run very close together with the latter moving a little faster; c, The first T of the EcoRI site of pBR322 is taken arbitrarily as base pair number 1. There are 4362 base pairs in this plasmid; d, A restriction map for EcoRI with λ DNA (sizes in kb). The two termini of λ have cohesive (or sticky) ends. After digestion with EcoRI, a 7th fragment of 25.4 kb may be observed as a result of annealing of the two end fragments through these cohesive ends. Heating the digest to 70°C followed by quick cooling will separate the two annealed fragments.

for *Bam*Hl (position 375) and *Sal* 1 (position to be determined). By carrying out a double digest with *Eco*RI and *Sal* 1 (track 5), two fragments will be obtained and their sizes can be determined. They should of course add up to the full length pBR322 size. It will be necessary to extrapolate the graph to measure one of the fragments. This is far from ideal, but will be sufficiently accurate for this experiment. It should be apparent that there are two positions in Fig. 2c at which the *Sal* 1 site could be located to generate these two fragments. A triple digestion of *Eco*RI, *Bam*Hl, and *Sal* 1 (track 6) will resolve which of the two locations is correct. Using the data obtained from tracks 5 and 6, draw a restriction enzyme map of pBR322.

The samples containing λ DNA digests have been heated at 70°C for 10 min. Since λ has cohesive termini, the two terminal fragments generated after digestion can stick together through hydrogen bonding. This leads to the appearance of a new fragment. Heating at 70°C destroys the hydrogen bonding. For the λ DNA mapping experiment, the positions of digestion by *Eco*RI are given in Fig. 2d, and the fragment sizes obtained in track 8 should be checked with these (track 7 contains undigested λ DNA). The sizes of the two fragments obtained with *Sal* 1 (track 9) should also be measured. There are actually two sites for *Sal* 1 within 500 bp of each other, but this internal fragment will not be seen with so little DNA applied to the gel. The *Sal* 1 sites can be treated as one site for this experiment. Again there are two possible positions on the map for the *Sal* 1 site, i.e., nearer to the left-hand end or nearer to the right in Fig. 2d. A double digestion with *Sal* 1 and *Eco*RI (track 10) should resolve the problem, with *Sal* 1 cleaving one of the *Eco*RI fragments. Again, using the data obtained from tracks 8, 9, and 10, along with the map given in Fig. 2d, draw a restriction map of λ DNA.

Reference

1.. Sutcliffe, J.G. (1979) Complete nucleotide sequence of the *Escherichia coli* plasmid pBR322. *Cold Spring Harbor Symp. Quant. Biol.* **43**, 77.

Chapter 5

Southern Blotting

Alan Hall

1. Introduction

The Southern blotting technique was developed by E. Southern in 1975 (*1*) to facilitate the identification of specific sequences within a complex mixture of DNA fragments. The technique itself has been modified over the years and is now an extremely sensitive and fast analytical technique used in almost every molecular biology laboratory.

The basic idea of the technique is to electrophorese a mixture of DNA fragments in an agarose gel. The mixture can range from a few fragments derived by restriction enzyme digestion of cloned DNA, to the million or so fragments generated by digestion of mammalian or plant genomic DNA. The DNA is then transferred, or "blotted," onto a piece of nitrocellulose filter paper as follows: the DNA is first denatured by soaking the gel in alkali; after neutralization the gel is placed on top of three sheets of Whatman 3MM filter paper soaked in high-salt buffer (*see* Fig. 1). A sheet of nitrocellulose is placed on top of the gel and dry paper tissues are placed on top of this. In this way, capillary action forces the buffer from the bottom filter papers through the gel and the nitrocellulose into the

a

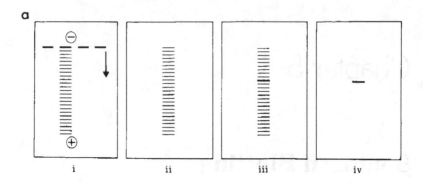

b

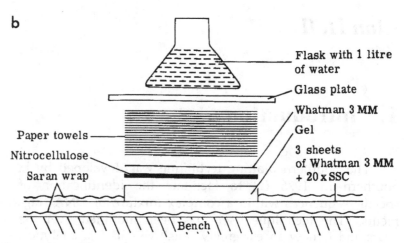

Flask with 1 litre
of water

Glass plate

Whatman 3 MM

Gel

3 sheets
of Whatman 3 MM
+ 20 x SSC

Paper towels

Nitrocellulose

Saran wrap

Bench

Fig. 1. Southern blotting procedure: a, A diagramatic
representation of the transfer process: (i) the DNA is seen in the
gel as a smear after ethidium bromide staining; (ii) a nitrocellulose replica is then made and the DNA is bound to the nitrocellulose; (iii) the radioactive probe finds the corresponding sequence
and binds to it. Nothing is actually seen on the filter in steps (ii)
and (iii), but if an autoradiograph of the filter is made, as in (iv),
then a band can be observed; b, The transfer protocol.

dry tissues. It carries along with it the DNA from the gel,
which becomes trapped on the nitrocellulose, and a replica
of the gel is obtained. The nitrocellulose is then heated for
2 h at 80°C, binding the DNA tightly (though not
covalently) to the filter.

It is now possible to identify sequences on the filter by
hybridization to a defined radioactive DNA or "probe."
The filter is usually incubated with radioactively labeled
DNA under conditions of salt and temperature that allow
hybridization of the probe to any homologous sequences

present on the filter. After washing the filter, hybridized material may be detected by autoradiography. For example, the globin genes present in human DNA can be detected by electrophoresing digested total human DNA in an agarose gel (*see* Fig. 1a, panel i), Southern blotting onto nitrocellulose (Fig. 1a, panel ii), and probing with a cloned radioactively labeled globin gene (Fig. 1a, panel iii). The probe will hybridize with fragments of DNA containing the globin gene, and the hybridized probe is then visualized by autoradiography (Fig. 1a, panel iv). This represents one out of the million or so fragments present and is usually about 10 pg of DNA (10^6 molecules of globin sequence). This is clearly a very sensitive technique and provides rapid analysis of genetic information.

Northern blotting was developed later and is based on the same principle as Southern blotting, except RNA is electrophoresed through the gel, transferred to nitrocellulose, and probed (2). In this case the level of any particular RNA species is not fixed (as with DNA) and will vary depending on the level of expression in the cell line from which the RNA was obtained.

Details concerning some of the steps of the Southern blotting technique are given below.

1.1. Transfer Step

The process of blotting is such that large fragments (>10 kb) are transferred to the nitrocellulose much more slowly than small fragments. In order to speed this up and ensure even transfer over the whole size range, the gel can be presoaked in dilute hydrochloric acid. This results in partial depurination of the DNA. When the gel is then soaked in alkali, primarily to denature the DNA to provide single strands for hybridizing to the probe, the DNA backbone is broken adjacent to depurinated residues. Large fragments are broken into small ones and they transfer quickly (approximately 3 h).

1.2. Preparation of Radioactive DNA

Obtaining highly radioactive probes is relatively easy using the nick translation procedure (3). The DNA probe is mixed with a small amount of DNAse-I, DNA polymerase I (pol 1), and one radioactive and three nonradioactive deoxyribonucleotides. The DNAse nicks the DNA, pol 1 removes the base to the 3′ side of the nick (Fig. 2), and then puts a base back on the 5′ side. As the radioactive nucleotide is incorporated, the fragment itself becomes labeled. In this way, high specific activities of 2×10^8 cpm/μg of DNA can be obtained.

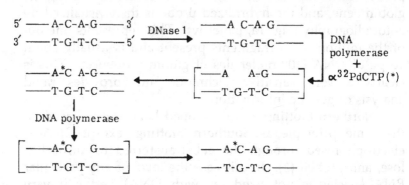

Fig. 2. Nick-translation reaction: DNAse 1 randomly introduces single-strand breaks into double-stranded DNA (nicks). DNA polymerase will then remove a base at the 3′ side of the nick using its $5' \rightarrow 3'$ exonuclease activity. Polymerase will then incorporate a nucleotide in the gap as shown. If the reaction medium contains dGTP, dATP, dTTP, and α^{32}P-labeled dCTP, then a radioactive phosphate group will be incorporated into the DNA along with dCTP. DNA polymerase can repeat this exonuclease/filling-in reaction along the DNA, hence the term nick translation.

1.3. Hybridization Step

The hybridization conditions are particularly critical. The speed with which the probe binds to the homologous sequence on the filter depends on the buffer composition and the temperature. All double-stranded DNAs have a characteristic melting temperature, the T_m, and it has been found empirically that the optimum temperature for hybridization is $T_m - 20°C$. In practice, hybridizations are often carried out at 68°C in 0.5M NaCl for convenience. Alternatively, formamide can be added to the hybridization mixture. The melting temperature of DNA is lowered by 0.7°C for every 1% formamide added, and conditions of 42°C, 50% formamide, are commonly used. Nitrocellulose does become more brittle at higher temperatures, so formamide addition is often preferred. It has been found that the speed of hybridization can be increased tenfold by adding 10% dextran sulfate to the solution.

1.4. Washing Step

Washing the filter must also be carefully controlled. For homologous probes, i.e., those identical at every position to the sequence being analyzed, the wash is usually 0.015M NaCl at 65°C. However, it may be possible to detect sequences only distantly related to the probe. In this case, the sequence homology is not complete, the T_m of the hybrid will be less, and the salt concentration in the wash must be increased and the temperature decreased.

The experiment described in this chapter will take 2.5 d to perform and will show all the main manipulations involved. On the first day, EcoRI digested λ DNA will be electrophoresed through an agarose gel and transferred to nitrocellulose overnight. The following day the filter will be baked. Radioactive λ DNA will be used to hybridize to the filter overnight. On the third day the filter can be washed and an autoradiograph exposed overnight. This can be developed any time. Students should work in pairs for this experiment, although the nick translation procedure used to make the radioactively labeled λ DNA will be done by a member of the staff and demonstrated to the class.

The gel apparatus described in this chapter is based on wicks rather than a submerged gel. This design is offered as an alternative to that described in Chapter 3, but submerged gels can be used if desired. To allow digestion and electrophoresis of λ DNA all in one day, a smaller agarose gel than that described in Chapter 3 is recommended. A minigel system, such as that described in Chapter 9, is suitable, but if only larger gel tanks are available, the gel will need to be run for a longer period—throughout the day or overnight.

2. Materials Provided

2.1. Electrophoresis

Agarose
10× Concentrated electrophoresis buffer
Deionized water
EcoRI digested λ DNA (0.1 mg/mL)
Agarose gel-loading buffer
Ethidium bromide solution
Holder and comb for gel
Electrophoresis equipment

2.2. Southern Blotting

Denaturation solution (0.5M NaOH, 1M NaCl)
Neutralization solution (0.5M Tris-HCl, pH 7.5, 3M NaCl)
20× SSC (3M NaCl, 0.3M sodium citrate)
Nitrocellulose sheet
Whatman 3MM filter paper
Paper towels
Saran wrap
80°C Oven (preferably a vacuum oven)
Trays for staining and soaking gels
Tweezers

2.3. Hybridization and Washing

Plastic bag and heat sealer
Prehybridization solution
Hybridization solution
42°C Waterbath
65°C Waterbath
2× SSC, including 0.1% SDS
0.1× SSC
Film holder and film for autoradiography
Access to darkroom and developing facilities

2.4. Nick Translation

A nick translation kit may be used, or the following:
Nucleotide/buffer solution
DNAse-I solution

DNA polymerase-I solution
$[\alpha - {}^{32}P]$ d CTP

The following are required, with or without the nick translation kit.

15°C Waterbath
Water saturated phenol
Sonicated salmon sperm carrier-DNA
G75 Sephadex and columns
Glass wool
TNE buffer
Microfuge tubes
Geiger counter
Radiation shield
Radioactive waste bin for solids and for liquids
Scintillation vial, scintillant, and access to scintillation counter

3. Protocol

3.1. Day 1

1. Set up a small agarose gel, as shown in Fig. 3a. Place the following in a conical flask and boil for 1 min:

    0.5 g Agarose
    5 mL 10× electrophoresis buffer
    45 mL water

    Allow the agarose solution to cool to hand-hot, then pour the molten gel into the mould and stand the comb about 2 cm from one of the two short sides, as shown in Fig. 3a. Allow to set.

2. To test the sensitivity of the Southern blotting procedure, different amounts of the λ/EcoRI digest can be loaded. Make two dilutions of the λ/EcoRI stock solution as follows:

    Stock solution: 500 ng/5 μL
    Dilution 1:   5 ng/5 μL. Add 1 μL stock solution to 99 μL water in a microfuge tube. Mix well
    Dilution 2:   50 pg/5 μL. Add 1 μL dilution 1 to 99 μL water. Mix well

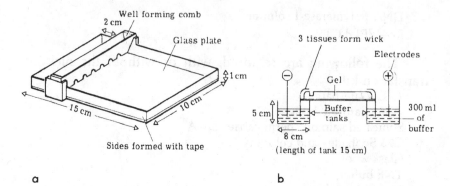

a   b

Fig. 3. Electrophoresis procedure: a, Tape is placed around all the sides of a glass plate as shown. Molten gel is poured into this mould and a plastic comb placed in the gel; b, After the gel has set, the comb and tape are removed and the glass plate is rested on two buffer tanks as shown. Three Kleenex tissues, wetted in buffer, are placed at either end to form wicks. The wells of the gel are filled with buffer and the samples for electrophoresis are layered in the well, thereby displacing the well buffer.

Next set up the following in three microfuge tubes:

	a	b	c
λ/*Eco*RI digest	5 μL	5 μL	5 μL
	stock soln	(dilution 1)	(dilution 2)
Deionized H_2O	25 μL	25 μL	25 μL
Agarose gel-loading buffer	20 μL	20 μL	20 μL

Note: *Volumes of deionized H_2O and loading buffer may be varied according to the size of the wells in the agarose gel.*

3. Set up the gel for electrophoresis as shown in Fig. 3b and fill all the wells with electrophoresis buffer. If a submerged gel is being used, assemble the gel according to the manufacturer's instructions. Using disposable pipet tips, load samples a, b, and c under buffer in the sample wells. Leave an empty lane between a and b, and b and c. Electrophorese at 75 mA for 2 h (current and time may vary according to the design of equipment used). Remove the gel and soak in 100 mL of 1 μg/mL ethidium bromide in the con-

tainer provided for 20 min. *Always wear gloves when handling ethidium bromide.* Destain for 10 min in water. Photograph the gel.

4. Place 200 mL denaturation solution into the container provided. Carefully remove the gel from its holder and place in the alkali solution. Agitate gently for 30 min. Carefully pour off the alkali and add 100 mL neutralization solution. Agitate gently for 15 min. Carefully pour off and replace with a fresh 100 mL neutralization solution. Leave another 15 min. Step 5 can be carried out while waiting for the final 15-min soak.

5. Measure the exact size of the gel and cut out the same size of nitrocellulose. Gloves should always be worn when handling the nitrocellulose to avoid getting grease marks on the filter. It is easier to cut the filter while it is still sandwiched between the protective sheets of paper. Lay the filter slowly on top of a solution of $2\times$ SSC. (Dilute $20\times$ SSC ten times with deionized water) so that the filter wets evenly. After a couple of minutes, lift the filter carefully with tweezers and turn it over to wet completely.

6. The gel is now ready to transfer. Place a sheet of Saran wrap (30×30 cm) onto a flat bench. Place a sheet of Whatman 3MM filter paper (15×20 cm) on top of the Saran wrap and soak with $20\times$ SSC. Lay a second sheet of filter paper on top of the first and again soak. Repeat a third time. Carefully lift out the gel from the neutralization solution and place in the center of the wet filter paper (*see* Fig. 1b). Make sure it is facing the way it was during the electrophoresis. Place Saran wrap all around the gel to cover the filter paper. The Saran wrap should completely cover the wet paper and be flush with the sides of the gel. This cuts down evaporation and prevents the buffer from bypassing the gel and directly entering any overhanging tissue paper. Gently roll a 10-mL pipet over the gel to squeeze out any air bubbles that may be trapped underneath. Lay the sheet of nitrocellulose on top of the gel, and roll again to

remove any air bubbles. Place a sheet of Whatman 3MM filter paper (or equivalent), the same
size as the gel, on top of the nitrocellulose, and
then stack about 2 in of dry paper towels on top
of this. Again the paper towels should be roughly
the same size as the gel. Place a suitable weight,
such as a flask containing 1 L of water, on top of
the sandwich to hold down the towels and leave
overnight.

3.2. Day 2

1. Remove the paper towels, and place the nitrocel-
    lulose filter in 100 mL of $2\times$ SSC to rinse for 30
    s. Blot dry with filter paper, place between two
    sheets of dry filter paper, and bake at 80°C for 2
    h in a vacuum oven.

2. During this time, the procedure for nick translat-
    ing DNA will be demonstrated (*see* section 3.4).

3. After baking, place the filter inside a plastic bag
    and seal with a commercial bag sealer on three
    sides. Add, through the open end, 5 mL
    prehybridization solution.   This solution will
    mimic the actual hybridization conditions. Any
    nonspecific binding of DNA to the filter will take
    place now, with the nonradioactive DNA in the
    prehybridization solution. Remove any large air
    bubbles and seal the fourth side, leaving about a
    1.5-in. gap from the filter on this side. Place in a
    42°C waterbath for 2 h.

4. Cut open the fourth side, lay the bag on top of a
    few sheets of tissue paper, and, by rolling a pipet
    firmly over the bag, squeeze out as much of the
    prehybridization solution as possible.

5. Add 4 mL hybridization solution. Samples of
    4×10^6 cpm of denatured nick-translated probe
    will have been added. *You must wear gloves at all
    times when handling radioactive material. Monitor the
    bench and work areas with the monitor and discard all
    radioactive waste in the specially marked receptacles.*
    Remove any large air bubbles and seal the bag

close up to the filter. Manipulate the liquid in the bag to mix well, and cover the whole surface of both sides of the filter with radioactive hybridization solution. Place in the 42°C waterbath and leave overnight.

3.3. Day 3

1. Stand a beaker containing 200 mL of 0.1× SSC in a 65°C waterbath to preheat. Wearing gloves, place the bag containing the filter on several sheets of tissue paper and cut on three sides. Peel out the filter and place in a tray containing 200 mL of 2× SSC and 0.1% SDS at room temperature. Agitate for 10 min. (Discard the bag and its contents into the radioactive waste bin.) Carefully pour off the first wash solution into the special radioactive waste sink, and replace with another 200 mL of 2× SSC and 0.1% SDS. Agitate for another 15 min again at room temperature.

2. Pour off the second wash and replace with 100 mL of 0.1× SSC at 65°C. Gently agitate the filter in a 65°C waterbath for 15 min. Pour off the third wash and repeat once more with 0.1× SSC at 65°C.

3. Blot the filter to remove excess liquid and wrap in Saran wrap. Place the covered filter in the special film holder and take it to the darkroom. Place a sheet of film on top of the filter, close the film holder, and leave in a light-tight container overnight. The following day the film may be developed.

3.4. Nick Translation Procedure

1. The simplest method for nick translation involves using a specially made kit available from Amersham International or New England Nuclear. If this is available, follow the manufacturer's instructions exactly for carrying out the reaaction with 10 μL/

λ DNA (0.1 mg/mL) up to the point of having incubated for 90 min at 15°C. Then go on to step 3, in this section.

2. Alternatively, the following three solutions can be made up (see Appendix II, Chapter 5.)

10× Concentrated nucleotide/buffer solution
DNAse-I solution
DNA polymerase-I solution

Place 50 μL of α³²PdCTP (300 Ci/mmol, 1 mCi/ mL) in a microfuge tube and lyophilize in a vacuum desiccator. *Gloves should be worn at all times and checked regularly for contamination. It is advisable to work behind a perspex screen and wear safety glasses.* (See Appendix I, Safe Handling of Radioactivity.) The dCTP should be frozen first with dry ice to prevent splashing during lyophilization. Redissolve the radioactive material in 80 μL of deionized, distilled water by vortexing. Add 10 μL of λ DNA (0.1 mg/mL) and 10 μL of the nucleotide/buffer solution. Add 1 μL of DNAse-I (0.1 μg/mL) and 1 μL of DNA polymerase-I, mix gently, and leave at 15°C for 90 min.

3. Meanwhile, set up a small column of G75 Sephadex that will be used to purify the nick-translated DNA away from unincorporated nucleotides. Cut off the long thin end of a Pasteur pipet and plug the neck with a small piece of glass wool (see Fig. 4). Add 0.5 mL of TNE buffer to the pipet, checking to make sure it flows through. Add G75 Sephadex slurry and pack to about 0.5 in. from the top of the pipet. Allow one column vol of 1× TNE to pass through, then stop the column from drying out by stretching a piece of parafilm over the top or using a clip as shown in Fig. 4. Label 12 microfuge tubes 1–12 and place the column and the tubes (in a suitable rack) behind the Perspex screen.

4. After 90 min incubation, the nick translation reaction should be vortexed with 100 μL of water-

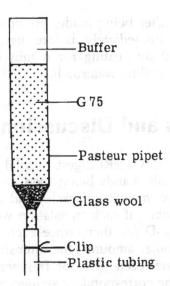

Fig. 4. Sephadex column to separate radioactively labeled DNA from unincorporated nucleotides after nick translation.

saturated phenol for 10 s and centrifuged in the microfuge for 1 min. Discard the bottom phenolic layer into the radioactive waste. Add 1 μL sonicated salmon sperm carrier-DNA (10 mg/mL) to the aqueous layer and mix. Allow the column to run until the buffer reservoir has just emptied, then layer the nick-translated material on top of the G75. Allow it to completely enter the Sephadex, then carefully fill up the reservoir with TNE. Collect 12 × 0.2 mL fractions and discard the column. By holding each fraction about 9 in. away from a geiger counter it should be possible to observe two peaks of radioactivity. The first peak should start about fraction 4–6 and is the nick-translated DNA. The second peak, about fraction 10, will be unincorporated nucleotides. Pool the fractions containing the DNA and discard the rest. Remove 1 μL of the pooled fractions, place in 10 mL liquid scintillant, and count in a scintillation counter. Calculate the incorporation of ^{32}P into the DNA. It should be around 3 × 10^8 cpm/μg. This is a sufficient supply of material for the experiment, and can be used up

to 2 wk after being made, providing it is kept at
−20°C. Immediately before use, the probe is
denatured by heating for 4 min in a microfuge
tube in a boiling waterbath.

4. Results and Discussion

Hybridization of λ/*EcoRI* digest to total λ-DNA should
result in all five *EcoRI* bands being seen on the autoradiograph. The relative intensity of each band depends on the
integrity of the probe. If nick translation were to result in
intact, full-length λ-DNA, then, since each *EcoRI* fragment
is present in equimolar amounts, the intensity of each band
should be the same; each mole of fragment hybridizes a
given amount of the corresponding sequence in the λ-DNA
probe, but since this sequence is attached to the rest of the
λ-DNA, all fragments will be equally radioactive. If after
nick translation very small (300—500 bp) fragments are generated, however, the amount of radioactivity hybridizing
will be proportional to the length of each fragment. The
size of the nick-translated product depends on the nick
translation conditions. High concentrations of DNAse-I
result in high levels of incorporation of radioactivity, but
also in small fragments.

The three lanes a, b, and c should give some idea of
the sensitivity of the blotting technique. In lane c, 50 pg of
λ-DNA have been applied and at least the larger fragments
should be visible in this track. It is relatively easy to detect
10 pg of a fragment using this technique. The human
genome contains enough DNA for about a million genes,
and to have one of these present at 10 pg means 10 μg of
DNA have to be loaded on the gel. This is easily achieved
using a 0.5−1-cm thick gel, in which a single copy gene
sequence can therefore be readily detected.

References

1. Southern, E. (1975) Detection of specific sequences among
    DNA fragments separated by gel electrophoresis. *J. Mol.
    Biol.* **98**, 503.

2. Thomas, P.S. (1980) Hybridization of denatured RNA and small DNA fragments transferred to nitrocellulose. *Proc. Natl. Acad. Sci. USA* **77**, 5201.

3. Rigby, P.W.J., Dieckmann, M., Rhodes, C., and Berg, P. (1977) Labeling deoxyribonucleic acid to high specific activity in vitro by nick-translation with DNA polymerase I. *J. Mol. Biol.* **113**, 237.

Chapter 6

Preparative Electrophoresis

Extraction of Nucleic Acids From Gels

Alan Hall

1. Introduction

Extraction of DNA or RNA fragments from agarose or acrylamide gels is an important tool in genetic engineering. A mixture of fragments can be separated by electrophoresis, and a particular fragment eluted and used for subsequent analysis or cloning.

1.1. Extraction From Acrylamide Gels

Acrylamide gels are used to separate fragments of relatively small size, usually less than 1 kb, and extraction of DNA from the gel is relatively straightforward. First the fragment of interest is identified in the gel after electrophoresis. Although ethidium bromide staining of DNA in acrylamide gels can be used, it is less sensitive than when used in agarose gels. Therefore, radioactive labeling of the fragment prior to electrophoresis is often preferred. The fragments may then be identified by autoradiography. A gel slice containing the DNA fragment of interest is cut out, crushed in an extraction buffer, and left agitating overnight. The gel material can be removed by filtration or centrifugation, and the DNA is precipitated from the supernatant with ethanol. Relatively high yields are usually obtained (>50%), but the yield drops markedly as the fragment size increases. This procedure can also be used for RNA, although in this case particular care must be taken to see that all solutions and equipment that come into contact with the material are free of ribonucleases.

1.2. Extraction From Agarose Gels

DNA fragments larger than 1 kb are separated on agarose gels. A multitude of methods have been reported for extraction of DNA from agarose gels, ranging from total dissolution of the gel in concentrated potassium iodide solutions, to the continuous elution of DNA off the end of the gel with a special apparatus for collecting fractions. However, two particularly useful techniques for eluting fragments have been widely used, both having the advantage of being simple and reproducible. First, the DNA can be electroeluted from the gel. This is carried out by removing a gel slice containing the fragment of interest and placing it inside a small piece of dialysis tubing that has been closed at one end. The bag is filled with buffer, closed at the other end, and immersed in electrophoresis buffer. A current is passed through the buffer and the DNA migrates out of the gel and accumulates on the inside of the dialysis membrane. The current is reversed for a minute to pull the DNA off the membrane, and the buffer inside the bag, now containing the DNA, can be removed leaving behind the gel.

The second technique is the one used in the experiment described in this chapter. This makes use of low-gelling-temperature agarose, a specially purified grade of agarose that once set, can be remelted by heating to 65°C. Furthermore, it will resolidify only below 32°C. The fragment is cut out in a gel slice that is then melted at 65°C, diluted with buffer at 65°C, and extracted with phenol at 65°C. Phenol extraction is a very common procedure in molecular biology. If water-saturated phenol is shaken with a DNA solution (usually at room temperature) and then centrifuged, two layers are formed; a bottom layer of phenol, and an aqueous layer on top. The DNA stays in the aqueous layer, but proteins are extracted into the phenol layer or are denatured and stay at the interface. This is the standard way of removing contaminating proteins or enzymes from DNA solutions.

In the case of gel extraction, the phenol and molten gel layer are shaken at 65°C (to prevent the gel resolidifying) and the agarose forms a white precipitate at the interface of the two layers. In this way the agarose can be removed from the solution and the DNA precipitated from the aqueous layer with ethanol.

This type of agarose gel is even more versatile; since the molten gel will not resolidify at 37°C, restriction enzyme digests can be carried out on DNA fragments still in the gel. Even more surprising, if two fragments are to be joined together (*see* Chapter 7 on DNA ligation), it is possible to melt two gel slices, each containing a different fragment, mix them together, and add the DNA-joining enzyme (DNA ligase). Although the gel is incubated at 15°C (the usual temperature for DNA ligation reactions), and therefore solidifies, the ligation reaction still works! This is a very important time-saving device; many different recombinants can be generated quickly using different fragments, without having to extract each one before joining.

Extraction of RNA from agarose gel is carried out in the same way as described for DNA, though once again all solutions must be absolutely free of ribonucleases.

In the experiment described in this chapter, λ DNA that has been digested with *Eco*RI will be run on a low-gelling-temperature agarose gel, and two of the five *Eco*RI fragments extracted. The experiment will take 1.5 d to perform. The first day will involve digesting λ DNA with *Eco*RI, electrophoresing on a low-gelling-temperature agarose gel, and phenol extracting two fragments. The DNA will be ethanol precipitated overnight. The next day the ethanol precipitate will be collected and an aliquot electrophoresed on a normal gel alongside the original λ DNA digest. The purity and yield of the extracted fragments may then be determined. The gel apparatus used is the same as that described in Chapter 5, but a submerged minigel apparatus can be used instead, if preferred (*see* Chapter 9).

2. Materials Provided

2.1. Solutions for Digestion

Deionized water
10× Concentrated EcoRI digestion buffer
λ DNA (0.2 mg/mL)
EcoRI (10 units/μL)

2.2. Electrophoresis

Low-gelling-temperature agarose
Normal agarose
Deionized water
10× Electrophoresis buffer
Agarose gel-loading buffer
Ethidium bromide solution
UV light box, face mask, and photographing equipment
Gel-forming equipment
Electrophoresis apparatus

2.3. Gel Extraction

Gel extraction buffer
Water-saturated phenol
Chloroform
5M NaCl
Absolute ethanol
TE buffer
Razor blades

3. Protocol

3.1. Day 1

A low-gelling-temperature agarose gel should be prepared first (this can be setting while the digestions are being carried out).

1. Weight out 0.5 g low-gelling-temperature agarose and place in a conical flask with 45 mL deionized water and 5 mL 10× concentrated electrophoresis

buffer. Boil for 1 min with occasional shaking. Allow the agarose solution to cool slightly then, preferably in a cold room, pour the molten gel into a gel mould, and place the comb about 2 cm from the top, as shown in Fig. 3a, Chapter 5. The comb should be resting just off the surface of the glass plate. Leave at 4°C until ready for use.

2. Add the following solutions to a 1.5-mL micro-fuge tube:

> 20 µL DNA solution (0.2 mg/mL)
> 51 µL Deionized, distilled water
> 8 µL 10× Concentrated EcoRI digestion buffer

Close the cap, mix gently for a few seconds, and spin for 10 s in the bench microfuge to bring the contents to the bottom of the tube. Add 1 µL of EcoRI (10 units/µL). Mix well and incubate in a 37°C waterbath for 1 h.

3. When the gel has set, remove the comb. Low-gelling-temperature agarose is not as mechanically strong as normal agarose and great care must be taken in removing the comb. This is best done while the gel is still at 4°C. Set up the gel for electrophoresis, as shown in Fig. 3b, Chapter 5 (if a submerged gel is being used, set up the equipment according to the manufacturer's instructions). Fill all the wells with electrophoresis buffer.

4. Add 20 µL agarose gel-loading buffer to the λ/EcoRI digestion. Mix well. Place in a 70°C waterbath for 10 min, then on ice for 5 min. Load 40 µL into each of two neighboring wells of the agarose gel, and electrophorese at 75 mA for 2 h (current and time may vary according to the design of equipment used). Save the remaining 20 µL (0.8 µg of λ DNA) for use on the following day. (Store at 4°C.)

5. Turn off the current, remove the gel, and place in a container with 100 mL of 1 µg/mL ethidium bromide. *Always wear gloves when handling ethidium bromide solutions.* After 30 min of gentle agitation,

carefully pour off the ethidium bromide solution and wash the gel with water for 10 min.

6. Label five microfuge tubes with number 1, and five tubes with number 2.

7. Place the gel on a UV light box. (*Wear a face mask at all times when the light is on to protect your eyes and skin.*) The five fragments of λ DNA, generated by *Eco*RI digestion, should be visible. Choose one of the fragments and remove two gel slices with a sharp blade, one from each track. Place the two slices into a tube marked 1. Pick a second fragment and again cut it out from both wells and place in tube 2. The slice should be cut so that as little excess agarose as possible is taken. The total volume of gel from the two tracks should be approximately 0.1 mL.

8. Place 4 mL gel extraction buffer and 4 mL water-saturated phenol in a 65°C waterbath to preheat. Place the two tubes containing the gel slices in a 65°C waterbath for 5 min. The gel should melt. Add 0.5 mL gel extraction buffer (preheated at 65°C) to both tubes. Vortex for 10 s and return to the 65°C waterbath for 5 min.

9. Wearing gloves and safety glasses, carefully add 0.5 mL phenol (65°C) to each tube. *Phenol is very corrosive; if spilled on skin, wash well with soap and water.* Vortex the tubes for 2 min. The solution will stay warm during this procedure. Centrifuge in the bench microfuge at room temperature for 4 min. Carefully remove the top aqueous layer from tube number 1 and place in a clean tube also marked 1. Repeat for tube 2. The agarose will be seen as a white precipitate at the interface. Try to leave as much of this behind as possible when removing the top layer.

10. Repeat step 9 two more times. The top layer should be prewarmed again to 65°C for 1 min before adding the phenol. The volume of the aqueous layer decreases during the three phenol extractions.

11. Add 0.4 mL chloroform to each of the aqueous phases. Vortex for 30 s and spin in the bench

microfuge for 2 min. Remove the top aqueous layer and place in a new tube leaving behind all the chloroform. Chloroform removes the small amount of phenol that is dissolved in the aqueous layer.

12. Measure the volume of the aqueous layer with a pipet, add 1/50 vol of 5M NaCl, and mix. Add 2 vol of ethanol and mix. Place at −20°C overnight. (In 0.1M NaCl and 66% ethanol, DNA will precipitate at low temperatures.)

3.2. Day 2

1. Prepare a small agarose gel in the same way as described in step 1. However, use the normal agarose this time. The gel can be setting while the DNA precipitate is being collected.

2. Spin the two ethanol precipitates in a bench microfuge (preferably in a 4°C room) for 15 min. Carefully remove the supernatant with a micropipet. The DNA should be pelleted at the bottom of the tube so care must be taken not to touch the bottom of the tube with the tip. Leave behind about 50 μL of supernatant. It is unlikely that the DNA pellet will be visible, especially if the smaller fragments are being isolated.

3. Close the cap and spin again for 30 s. This will enable *all* the remaining supernatant to be removed. If this is done carefully it should not be necessary to dry the pellet. Add 30 μL of TE buffer to each tube and vortex for 1 min to dissolve the DNA fragments. Add 20 μL of agarose gel-loading buffer.

4. When the gel has set, assemble the gel for electrophoresis as shown in Fig. 3b, Chapter 5. Load as follows:

> track 1—λ/EcoRI digest (20 μL from section 3.1, step 4)
> track 2—fragment 1 extracted from gel (50 μL)
> track 3—fragment 2 extracted from gel (50 μL)

Electrophorese for 1 h at 75 mA and stain the gel
with ethidium bromide (*see* section 3.1, step 5).
Place on a UV light box and photograph.

4. Results and Discussion

It will be possible to estimate the purity of the isolated
fragments from looking at tracks 2 and 3. When fragments
are close together on the low-gelling-temperature agarose,
the chances of contamination are obviously greater. The
yield of fragments can be roughly estimated by comparing
the intensity of the isolated fragment with the corresponding fragment in track 1. The latter is 20% of the starting
material. Yields should be of the order of 50%; therefore
the intensity of the bands in tracks 2 and 3 should be twice
that of the corresponding band in track 1.

Chapter 7

DNA Ligation and *Escherichia coli* Transformation

Alan Hall

1. Introduction

The generation of recombinant clones usually involves joining together covalently (also called ligating) two or more restriction fragments. This can be done with the enzyme DNA ligase. There are two major sources of the enzyme: *Escherichia coli* DNA ligase, which requires Mg^{2+} and NAD^+ as cofactors and is less commonly used, and T4 DNA ligase produced by the *E. coli* phage T4, which requires Mg^{2+} and ATP as cofactors. The in vivo function of both these enzymes is to repair nicks in double-stranded DNA, provided there is a 5′-phosphate group and a 3′-hydroxyl group (*see* Fig. 1). Since many restriction enzymes generate sticky ends (Fig. 1B), ligase will form two covalent bonds to give a single double-stranded fragment. This can be thought of as repairing two nicks. The reaction is usually carried out at 15°C to optimize both the enzymatic activity and the percentage of sticky ends actually in the annealed form. Since this only involves hydrogen bonding between a few base pairs, the dissociation temperature is very low, near 5°C. Unlike *E. coli* DNA ligase, T4 DNA

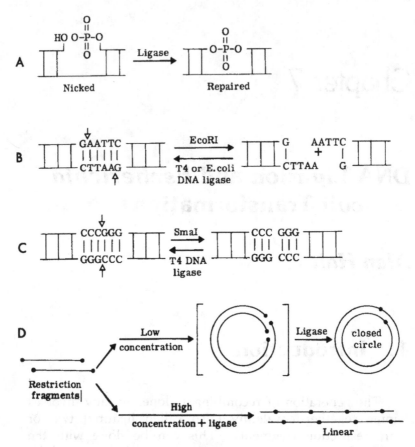

Fig. 1. Enzymatic properties of DNA ligases: A, DNA ligases repair single-stranded breaks (nicks) in a double-stranded DNA; B, DNA ligase can rejoin two DNA fragments that have cohesive (or sticky) ends, for example as generated by the restriction enzyme *EcoRI*; C, T4 ligase will rejoin two DNA fragments that have blunt ends, for example as generated by the restriction enzyme, *SmaI*; D, The concentration of DNA fragments in a ligation reaction can have a large influence on the outcome. At low concentrations intramolecular circularization will predominate, whereas at high concentration, intermolecular oligomerization will be favored.

ligase will also covalently link two blunt-ended molecules (Fig. 1C), and this reaction can be used to join ends generated by enzymes that do not give sticky ends.

The other major variable in ligation reactions is the DNA concentration. At low DNA concentrations (5 µg/

mL) the unimolecular circularization reaction is favored, whereas at high concentrations (100 μg/mL), polymerization predominates (Fig. 1D). Cloning experiments, for example the insertion of a foreign EcoRI fragment into the EcoRI site of pBR322 (Fig. 2A), are usually carried out at relatively high DNA concentrations (20 μg/mL) and with a molar ratio of insert DNA to vector DNA of around 3:1. Even so, only 10–20% of molecules obtained will actually be recombinants, the rest will be recircularized pBR322. There are ways to increase the number of recombinants. For example, if pBR322 is cleaved with two different enzymes, say EcoRI and BamHl (Fig. 2B), and the resulting large fragment isolated from a gel, this will not recircularize with DNA ligase. However, if foreign DNA with EcoRI and BamHl ends is added along with ligase, then almost all the molecules obtained will be recombinants. Another way of optimizing recombinants is to make use of the enzyme phosphatase. This enzyme will remove phosphate groups from the ends of DNA molecules. Since DNA ligase requires a 5'-phosphate group for ligation to occur, EcoRI-linearized pBR322 cannot be recircularized with ligase if it is first treated with phosphatase. A foreign EcoRI fragment (that still has its 5'-phosphate groups) can then be added along with DNA ligase, and once again almost all the molecules obtained will be recombinants (Fig. 2C).

DNA ligation is usually followed by transformation of E. coli cells. The products of DNA ligation are complex, but by putting the mixture back into E. coli, those molecules that are still biologically active will be selected. Biological activity is usually determined by an antibiotic drug resistance gene carried on the plasmid. For instance, pBR322 confers resistance to both ampicillin and tetracycline when present in E. coli cells. It has been found that if E. coli cells are preincubated with $CaCl_2$ for 0.5 h, they are then susceptible to DNA uptake. After $CaCl_2$ treatment, the cells are said to be competent. The efficiency with which they take up DNA varies enormously from E. coli strain to E. coli strain, and depends on exactly how the cells are grown and treated. DNA is added to the competent cells and after incubation the cells are put onto agar plates containing the antibiotic ampicillin. Only an E. coli cell that has taken up a molecule of pBR322 can grow on the medium to form a

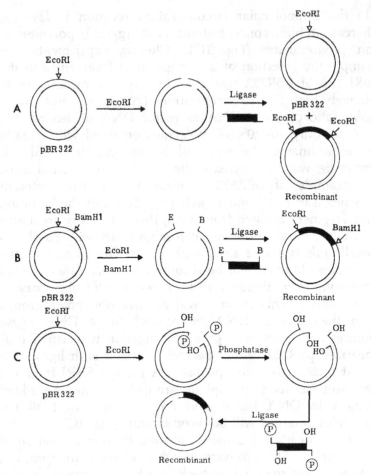

Fig. 2. Cloning of foreign DNA fragments: A, Digestion of pBR322 with *Eco*RI yields a linear molecule. Foreign DNA (with *Eco*RI ends) can be cloned into this site, but under normal ligation conditions recircularized plasmid is the predominant product and only 10–20% of the molecules will be recombinants; B, If pBR322 is cleaved with two different enzymes (*Eco*RI and *Bam*Hl in this case) and the large fragment purified away from the small fragment, then foreign DNA (which must have an *Eco*RI site at one end and a *Bam*Hl site at the other) can be efficiently cloned into this. Over 90% of the molecules are likely to be recombinants; C, If pBR322 is cleaved with *Eco*RI and then treated with the enzyme phosphatase, the linear plasmid can no longer be recircularized with ligase. However, if a foreign fragment (with *Eco*RI ends carrying intact phosphate groups) is added then after ligation, over 90% of the molecules will be recombinants. The remaining nick in both strands is repaired efficiently in vivo after transformation of *E. coli* cells.

visible colony of cells after 12 h incubation. With the procedure described here for making competent cells from the *E. coli* strain HB101, about 5×10^5 ampicillin-resistant colonies should be obtained per microgram of pBR322 DNA added.

The *EcoRI* site in pBR322 is not within either the ampicillin or the tetracycline genes, and recombinants formed by insertion into this site are still resistant to both antibiotics. However, inactivation of drug resistance can be exploited with pBR322. For example, the single site for *PstI* is within the ampicillin gene and insertion of DNA into this position will inactivate the gene. In this case the ligation mixture is used to transform cells that are then spread on tetracycline plates. The percentage of recombinants can be determined by testing a number of the tetracycline-resistant colonies for ampicillin sensitivity.

The experiment described below will entail linearizing pBR322 with *EcoRI*. The *EcoRI* will be heat inactivated and the DNA religated with DNA ligase. The efficiency of these reactions will be tested by comparing the transforming efficiency of competent cells with pBR322, *EcoRI* digested pBR322, and religated pBR322. The experiment will take 1 d to perform. Transformed cells will then be plated out and left overnight to grow. The plates can be transferred to 4°C the next morning until ready for scoring.

2. Materials Provided

2.1. *Solutions for Digestion*

pBR322 (0.2 mg/mL)
Deionized, distilled water
10× Concentrated *EcoRI* digestion buffer
EcoRI (2 units/µL)
Microfuge tubes

2.2. *Solutions for Ligation*

10× Concentrated ligase buffer
T4 DNA ligase (2 units/µL)

2.3. Preparation of Competent Cells and Transformation

Overnight suspension of HB101 cells
L broth
0.1M CaCl$_2$
7 Dry ampicillin plates
Sterile, 250-mL conical flask
Sterile centrifuge tubes for 50 mL of cells
4 Sterile test tubes

3. Protocol

1. Remove 2 mL of the overnight suspension of HB101 cells and inoculate into 50 mL of L broth in a 250-mL conical flask. This should be done under sterile conditions. Put the flask in a 37°C shaker. The cells will be used later to make competent cells. They should be monitored occasionally (after 1 h and then every half-hour) by measuring the absorbance at 650 nm in the spectrophotometer. The procedure will be demonstrated. The cells should be put on ice when they reach A_{650} 0.4–0.6.

2. After setting up the cells to incubate, begin the pBR322 EcoRI digestion. Add the following solutions to a 1.5-mL microfuge tube:

    40 μL Deionized, distilled water
    5 μL 10× Concentrated EcoRI digestion buffer
    5 μL pBR322 (0.2 mg/mL)

    Mix and centrifuge briefly (10 s) to bring the contents to the bottom of the tube. Add 1 μL of EcoRI (2 units/μL). Mix gently, centrifuge briefly, and incubate in a 37°C waterbath for 1 h.

3. Heat inactivate the EcoRI at 68°C for 10 min. Cool on ice 5 min.

4. Set up the following reactions in three labeled microfuge tubes.

	a	b	c
pBR322 (0.2 mg/mL)	1 µL	—	—
pBR322 Digested with EcoRI (step 3, 0.02 mg/mL)	—	10 µL	10 µL
Deionized distilled H$_2$O	9 µL	—	—
10× Concentrated ligase buffer	1 µL	1 µL	1 µL
T4 DNA ligase (2 units/µL)	—	—	1 µL

Mix gently and briefly centrifuge (10 s). Incubate at 15°C for 2 h.

5. The HB101 culture should have reached the correct absorbance during this time and should be sitting on ice. After sitting on ice for 10 min, the cells are transferred to a sterile centrifuge tube also sitting on ice, and then centrifuged at 4000 rpm for 20 min in the cold.

    Note: *From now on all procedures are carried out at 4°C or on ice.*

6. Pour off the supernatant and add 20 mL ice-cold 0.1M CaCl$_2$. Resuspend the cells carefully—avoid vortexing unless required for complete suspension. Leave on ice for 20 min.

7. Spin down the cells again, as in step 5. This time pour off the supernatant, then remove the last traces of liquid with a Pasteur pipet. Resuspend the cells in 0.5 mL 0.1M CaCl$_2$. Again, use as gentle conditions as possible. The cells are now "competent." They can remain on ice for up to 1 d or be used immediately. It is even possible to store them at −70°C for several months if 15% glycerol is added along with the CaCl$_2$, but this will not be necessary here.

8. Place the three tubes, a, b, and c from step 4, on ice for 2 min. Add 0.1 mL of competent cells to each and mix well by inversion and "flicking" the bottom of the microfuge tube. Leave on ice for 20 min. Add 0.1 mL of competent cells to an empty microfuge tube, d.

9. Place the tubes in a water bath at 42°C for 60 s *only*, and then back onto ice for 2 min. This heat shock appears to increase the efficiency of DNA

uptake. Add 1 mL of L broth to each of four sterile test tubes marked a–d and add the contents of the four microfuge tubes to each one. Shake gently at 37°C for 1 h. This time period allows the plasmid DNA that has been taken up to replicate and express its drug-resistance functions. Mark seven ampicillin L-agar plates a.1, a.2; b.1, b.2; c.1, c.2; and d, and spread the transformed cells as follows:

a.1	100 μL of L broth	10 μL, tube a
a.2	—	100 μL, tube a
b.1	100 μL of L broth	10 μL, tube b
b.2	—	100 μL, tube b
c.1	100 μL of L broth	10 μL, tube c
c.2	—	100 μL, tube c
d	—	100 μL, tube d

The cells are placed on the surface of the plate, then spread evenly over the surface. The best way is to use a triangle made from a long Pasteur pipet as shown in Fig. 3. Use a new spreader for

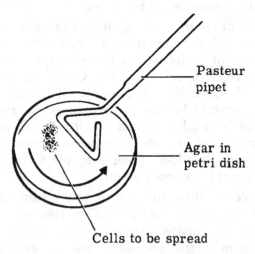

Fig. 3. Protocol for spreading plates: A spreader can be made from a long Pasteur pipet. If the pipet is held over a Bunsen flame for a few seconds, a triangular end can be easily made. After allowing enough time for cooling, this spreader can be used to smear the aliquot of cells over the surface of the agar. The Petri dish is turned in a circular motion with one hand and the spreader used with the other.

each plate. The plates are then incubated overnight in a 37°C oven.

11. The following day the plates can be scored or
placed at 4°C until it is convenient to analyze
them.

4. Results and Discussion

The number of ampicillin-resistant colonies obtained
should be counted. Use whichever plate from the two for
each experiment that has a reasonable number of colonies
visible (50 or so). Plate d of course received cells with no
DNA and this should be clean. Plate a will give the overall
efficiency of the uptake of DNA by the competent cells.
Since 0.2 μg of pBR322 was used in the transformation and
the transformed cells were then diluted up to 1 mL, the
efficiency, usually expressed as transformants/μg of DNA,
can be calculated. Plate b will give the "cut back"
efficiency of *Eco*RI digestion. This is usually of the order of
99% unless special steps are taken to increase it. For
instance, the linearized pBR322 could have been run out on
an agarose gel to separate the 1% of molecules that have
not been fully cleaved. This procedure is normally important for research experiments. Plate c shows the effect of
ligation, and the number of colonies obtained here should
be about 20–50% of the control experiment a. Recirculari-
zation is a very efficient process. Calculate the percentage
of the total pBR322 molecules that are taken up from the
DNA added to the cells in experiment a (the molecular
weight of pBR322 is 2.6 $\times$ 10^6 daltons). It will be clear that
many molecules are left behind. This is not so important
here, but in some cloning experiments, for instance cDNA
cloning, obtaining large numbers of clones can be difficult
and the efficiency of rescuing DNA molecules should be as
high as possible.

Chapter 8

Detection of Plasmid by Colony Hybridization

Elliot B. Gingold

1. Introduction

This experiment introduces colony hybridization, a technique central to genetic engineering technology. Why this technique is so important can be seen by considering the following problem. A scientific group wishes to clone in bacteria a particular human chromosomal gene. There would be no difficulty obtaining human DNA, cutting it with a restriction endonuclease, and ligating the many fragments obtained into vector molecules. This ligation mix could then be used to transform E. *coli* cells and plasmid-carrying colonies could be selected. The problem is that of identifying the clone carrying the required gene. Human chromosomal genes are most unlikely to be expressed in bacteria, and even less likely to give the colony a recognizable new phenotype. So how is it possible to select the one required clone from perhaps 100,000 others? By directly probing for the required DNA segment, colony hybridization provides a solution to this problem.

The principle behind nucleic acid hybridization is quite straightforward. If DNA is either heated beyond a critical temperature or exposed to alkaline conditions, it will denature; that is, the double helix will separate into single strands. When the sample is then returned to normal conditions, it will slowly reanneal to form a double helix. However, this can only happen when a molecule finds a partner with a complementary base sequence. This may be the strand it was originally associated with, a second DNA molecule with the same sequence, or even a complementary RNA molecule. Thus it is possible to test a given nucleic acid sample for homology with a second nucleic acid preparation.

In most hybridization experiments, the DNA sample to be treated is first denatured and then immobilized by binding to a nitrocellulose filter while still in the single-stranded state. The test nucleic acid, or *probe*, is radiolabeled and, if DNA, also denatured. The filter is then soaked in a solution containing this probe. If sequences complementary to that of the probe are found among the single-stranded DNA bound to the filter, double helices will form and the probe itself will become stably bound to the filter. If no such sequences exist, the probe can simply be washed off. Thus, homology between the probe and the bound DNA can be detected quite simply by examining the filter for radiolabel. This is usually done by exposing the filter to an X-ray film, a process known as autoradiography (*see* Appendix I, "Autoradiography").

When one is searching through a library of bacterial clones for a particular human gene, a suitable probe would be the mRNA molecule made by the gene. This will, of course, be complementary to the coding region of the chromosomal gene and thus form hybrid helices with it. It is sometimes possible to isolate an mRNA species from specialized cells producing a particular protein in excess; for example, β-globin and α-globin mRNA were easily purified from erythrocytes. Nonetheless, even in the best systems only a small amount of mRNA can be obtained. Hence it is far more usual to convert this mRNA into cDNA (complementary DNA) with the viral enzyme reverse transcriptase, insert the cDNA into a plasmid, and by bacterial transformation, cell growth, and subsequent plasmid isolation, obtain large quantities of the sequence in the DNA form. This cDNA can then be radiolabeled, denatured, and used as a probe for the chromosomal gene.

In theory, it would be possible to test the DNA from each bacterial clone for hybridization with the probe separately, but such a search would be clearly laborious. Colony hybridization provides a method by which the DNA from tens of thousands of clones can be tested at one time on a single filter. This is achieved by growing the bacterial colonies on the surface of a nitrocellulose filter placed on a nutrient agar plate. As well as providing a good physical support for the colonies, the filter allows for unimpeded transport of all nutrients; hence, well-formed colonies are produced. After forming a replica of these colonies on a second filter, they can be lysed and the DNA denatured by alkaline treatment. Baking the filters in a vacuum oven will then firmly bind the DNA from each colony to the position on the filter originally occupied by the colony itself. When this filter is exposed to the probe, it will only hybridize with the DNA from colonies with homologous sequences. In other words, the probe will only bind to the filter at the positions on the filter where the complementary chromosomal gene is found and hence originally occupied by a bacterial clone carrying this gene. After any such positions are identified by autoradiography, reference back to the second, replica filter will enable isolation of the required clone.

The present exercise is designed to provide an introduction to the technique of colony hybridization, and as such the task has been somewhat simplified compared to that already described. Instead of tens of thousands of colonies, a filter onto which nine bacterial clones have been picked will be used. Some of these contain the plasmid pBR322, others do not. With the help of a radiolabeled pBR322 probe, the task will be to identify the plasmid-carrying clones. Of course, in this case it would be easier to do this by simply testing each clone for antibiotic resistance! Nonetheless, this experiment provides a good introduction to the techniques involved and the protocol can be used, unmodified, for the more genuine task of screening colonies for an unexpressed DNA sequence with a suitable probe.

The method itself is based closely on that of Hanahan and Meselson (1). The full procedure takes 3 d to the point of setting up the autoradiograph and a further visit to the laboratory to develop the film and observe the result. Many of the steps, however, involve only brief periods of work and the exercise thus is best run as part of a block of practical classes rather than as a single, stand-alone effort.

Students are provided with nine patches of bacteria picked onto a nitrocellulose filter and grown on a nutrient plate. On the first day the filters are simply transferred to a chloramphenicol plate. This antibiotic will stop growth of cells and prevent bacterial chromosomal DNA synthesis, but allow the plasmid to continue replicating. Thus after another period of incubation a significant amplification of any plasmid DNA present will be achieved. This, of course, is of great assistance in the detection of the plasmid (or indeed any plasmid-carried inserts).

On the second day the colonies are lysed by NaOH treatment and the DNA baked onto the filter in a vacuum oven. The filters are then immersed in a solution containing the denatured probe and left at 68°C overnight. This temperature is chosen because it is low enough to allow renaturing of DNA, yet high enough to allow reasonable kinetics. Finally, on the third day the filters are extensively washed, dried, and the autoradiographs prepared. Depending on the nature of the radioisotope used, the film will be ready to develop between 1 d and 2 wk after it is set up.

Materials Provided

2.1. Cultures

Nine E. coli isolates on a Millipore HATF filter
One LB + chloramphenicol plate

2.2. Solutions

0.5M NaOH
1M Tris, pH 8
1M Tris, 1.5M MaCl, pH 8
Hybridization buffer
pBR322 Probe (in dilute salt buffer)
Washing buffer
3 mM Tris base
PPO (2,5-diphenyloxazole) in ether (if [^{3}H]-label is used)
Radioactive ink

2.3. Equipment

Millipore tweezers
Whatman 3MM paper
Plastic tray
Vacuum oven
Petri dish
Tape
Plastic bag
Large beaker or glass dish
68°C Oven
Autoradiography equipment (*see* Appendix I)

3. Protocol

The filters that the bacterial isolates are grown on are Millipore HATF. Before use they are first floated on distilled water and then submerged to ensure even wetting. The filters are then packaged between Whatman 3MM filters and sterilized by autoclaving. Each filter is placed onto a moist LB plate and samples of the nine isolates picked onto the filter in a known order. A pin hole (or group of holes) at one end of the filter is used for orientation.

The plates are incubated overnight at 37°C prior to the practical class and each isolate will have grown into a small patch culture. It is at this point that a replica could be made by pressing the filter onto a second filter. Details of this procedure can be found in Hanahan and Meselson (1). In this exercise, such a replication is not necessary because the original source of the nine isolates is available.

3.1. Chloramphenicol Amplification

1. Sterilize the tips of a pair of Millipore tweezers by soaking in alcohol and then passing through a flame. Use these tweezers to transfer the filter from the LB plate onto an LB + chloramphenicol plate, the colonies remaining face up. Check that the filter is in contact with the plate over the total area and that no air bubbles are present.

2. Incubate the plates overnight at 37°C.

3.2. Lysis and Filter Binding

1. Cut a sheet of Whatman 3MM paper to give squares slightly larger than the diameter of the filter. In a plastic tray, stack three of these squares.

2. Saturate the stack with 0.5M NaOH. Place the filter on top (colonies upward) and leave for 5 min. The lysis of the cells will normally produce a noticeable change of appearance.

3. Blot the filter on a paper towel (from below), then repeat the previous step (step 2) with fresh 0.5M NaOH and new filters.

4. Repeat this procedure twice more, using 1M Tris, pH 8, to saturate the papers.

5. Finally, repeat the procedure once using 1M Tris, 1.5M NaCl, pH 8.

6. Leave the filters in air until they appear nearly dry.

7. Bake the filter in a vacuum oven for 2 h at 60°C.

3.3. Hybridization

1. A probe consisting of radiolabeled pBR322 is prepared by nick translation (see Appendix II, Chapter 8). You will be informed whether the label used was ^{3}H or ^{32}P. The probe is made up in a dilute salt solution, and will be boiled just before use to denature the DNA.

2. In all subsequent stages, remember that radioactive materials are being handled. There is a guide to working with radioactivity in Appendix I, at the end of this book. Read this and follow the instructions carefully.

3. Place 1 mL of the hybridization buffer in a plastic Petri dish, labeled with your name. Add 0.1 mL of the probe solution.

4. Using Millipore tweezers, place the filter face down in the Petri dish. When fully wet, invert the filter.

5. Seal the Petri dish with tape and place in plastic bag. A group of Petri dishes will be sealed in each bag.

6. The plates will be incubated overnight at 68°C.

3.4. Washing

1. The washing buffer is preheated to 68°C and all washes using it are performed at this temperature. Remember that all used solutions are to be discarded into the containers provided, since these are radioactive after contact with the filter.

2. Carefully remove the tape from your plate. Using Millipore tweezers, transfer the filter to a large beaker containing at least 200 mL of the prewarmed washing buffer. Provided filters can be identified (e.g., by the pattern of marker holes), it is possible to include more than one in the same dish. If this is done, however, the quantity of buffer must be proportionately increased. The hybridization buffer containing the probe will be disposed of by the demonstrators.

3. Place the beaker and filters at 68°C.

4. Continue the wash for 3–4 h with three changes of buffer.

5. The filters are then washed in 3 mM Tris base at room temperature. Wash for 1.5 h with two changes of the solution.

6. Dry the filter for 30 min at 80°C in the vacuum oven.

3.5. Autoradiography

1. If ^3H has been used to label the probe, place the dried filter in a glass dish and wet it evenly with 12 mL of PPO in ether. *Since PPO is a carcinogen, wear rubber gloves.* Allow the ether to evaporate.

2. Mark the filter with radioactive ink. This is essential so that individual autoradiograms can be identified and the colonies correctly orientated.

3. The techniques and procedures for autoradiography are described in Appendix I at the end of this book. If ^{3}H is used, it will take 2 wk until the photographs are ready to develop, but with ^{32}P an exposure of less than 24 h will be adequate.

4. Results and Discussion

This exercise will normally produce clear results with dark images at the points at which plasmid-carrying colonies have been positioned and, at the worst, very faint darkening at the positions of other colonies. It is usual for the images to be more intensely darkened around their edges. This appears to be a direct result of chloramphenicol amplification working only on growing cells. Growth in patch cultures will always be greatest at the edges rather than in the central region of a patch.

Reference

1. Hanahan, D. and Meselson, M. (1980) Plasmid screening at high colony density. *Gene* 10, 63–67.

Chapter 9

Microisolation and Microanalysis of Plasmid DNA

Elliot B. Gingold

1. Introduction

This exercise provides an introduction to the important techniques of microscale DNA isolation and analysis. Techniques such as these are central to modern molecular biology and provide an interesting contrast to the more traditional large-scale biochemical methodologies. The experiment itself will fit into a 3-h practical session and requires no expensive laboratory equipment.

Elsewhere in this book you will find experiments involving large-scale isolation of plasmid DNA (Chapter 2) and its analysis on agarose gels (Chapter 3). It is obvious, however, that such methods would become too laborious if plasmid DNA from a range of clones needed to be screened. Such a situation will frequently arise when the products of a ligation reaction are used to transform a bacterial culture. For example, an investigator may wish to clone a particular fragment from one plasmid into a second plasmid. Following treatment of the donor plasmid with an appropriate enzyme to cut out the desired sequence and the cutting open of the recipient plasmid, a ligation would be attempted. It is most unlikely, however, that only the required product would be found in the ligation mix; other combinations of the various fragments would form as well. Thus, when the ligation mix is used to transform the bacteria, it is necessary to examine a number of clones to find the desired recombined plasmid. The technique used in this experiment provides a rapid method of screening a range of bacterial clones and estimating the size of plasmids they carry.

In this particular exercise you will be provided with three streak cultures of bacteria growing on a nutrient plate. The three clones will be labeled A, B, and C; one clone carries no plasmid, one clone carries pBR322, and the third clone carries a pBR322 derivative, including an insert.

Your aim will be to distinguish between the three clones. Although the clones are presented to you as streak cultures, the method is exactly the same as that used when analyzing random colonies from a spread plate.

The plasmid DNA isolation involves lysis of the cells in a mixture containing lysozyme and a non-ionic detergent (Triton X-100), followed by heat treatment to denature protein and chromosomal DNA. The resulting insoluble matrix is removed, after cooling, by centrifugation. The plasmid DNA is then precipitated from the supernatant by isopropanol and redissolved in a suitable buffer. Because of the small-scale nature of the work, the entire procedure is carried out in microfuge tubes. A rapid indication of plasmid size is obtained by running the DNA in a minigel apparatus. The DNA obtained may also be further analyzed by subjecting it to restriction-enzyme digestion, if time is available.

2.   Materials Provided

2.1.   Cultures

LB plate with strains A, B, and C

2.2. Solutions

Triton mix
2 mg/mL Lysozyme in Triton mix
Isopropanol (freezer cooled)
TE buffer
Gel-running buffer
0.8% Molten agarose in gel-running buffer
1 μg/mL Ethidium bromide in small trays
pBR322 Standard
Loading buffer

2.3. Equipment

1.5-mL Sterile microfuge tubes
Zinc oxide tape
Microfuge
Boiling waterbath
Micropipets (0–20 μL, 0–200 μL) and sterile tips
Minigel apparatus and power pack
–20°C Freezer
Transilluminator (or UV lamp)
UV opaque glasses

3. Protocol

This exercise consists of two parts—DNA isolation and minigel analysis. To enable the gel to set, it should be prepared by one member of the group while the DNA isolation is in progress.

3.1. DNA Isolation

1. You are provided with a plate on which bacterial isolates A, B, and C are streaked.
2. Measure 25 μL of Triton mix into each of three 1.5-mL microfuge tubes using a micropipet. The tubes should be labeled A, B, and C using a pen whose marking will not come off in a boiling waterbath.

3. From each bacterial streak, A, B, and C, collect a loopful of the bacteria and resuspend the cells in the Triton mix. It is essential to complete resuspension because any lumps will badly reduce the quality of the result.

4. To each tube add 2.5 μL of 2 mg/mL lysozyme (made up in Triton mix). Immediately seal the tubes with tape and place in a boiling waterbath for 3 min. During this period the cells will lyse, proteins will denature, and the DNA will "melt."

5. Remove the tubes from the waterbath and cool in ice for 5 min. Although most of the chromosomal DNA will be unable to reanneal and remain with the cell debris in an insoluble matrix, the closed circular plasmid should be able to renature and remain soluble.

6. Centrifuge the tubes for 15 min in a microfuge. This should pellet the cell debris, denatured protein, and chromosomal DNA matrix, though most of the plasmid DNA will remain in solution.

7. Carefully remove the supernatants into fresh tubes labeled A, B, and C using a micropipet. Take care not to disturb the pellets.

8. Add an equal volume of isopropanol to each supernatant and place the tubes at −20°C for 30 min.

9. Centrifuge the tubes for 15 min in the microfuge.

10. Decant the alcohol and drain dry. Take care that all the alcohol has been removed before proceeding. The pellet should be barely visible.

11. Resuspend the DNA in 4 μL of TE.

3.2. Minigel Analysis

1. During the DNA isolation procedure, the minigel is prepared. A number of designs of minigel apparatus are available commercially and each is supplied with its own instructions as to the manner of pouring gels, the volume of agarose gel to be used, and the quantity of buffer to be added. Following the procedure recommended in

the manufacturers instructions, pour a 0.8%
agarose gel using the molten agarose solution provided. It is desirable to allow at least 1 h for the
gel to set before loading the samples.

2. Label an additional microfuge tube D and add to
this 4 μL of pBR322 standard.

3. To each tube, add 1 μL of loading buffer. Mix
well and then centrifuge for 1 s to settle contents
at the bottom of the tubes.

4. Load samples A, B, C, and D onto the gel. A
micropipet is a convenient instrument for getting
the samples into the wells. The gel is run at 100 V
for 1 h or until the blue dye reaches the end of
the gel. If time is short, one can still achieve adequate separation with a briefer run.

5. Transfer the gel to a small tray containing 1 μg/
mL ethidium bromide. *Wear rubber gloves, since
ethidium bromide is a potential carcinogen.* Because
the gel is thin, 10 min staining should be adequate.

6. The gel can now be observed using either a transilluminator or direct UV illumination. Details of
viewing and photography are given in Chapter 3,
and should be followed, *but remember to avoid any
direct exposure to UV and to wear protective glasses at
all times.*

7. Note the appearance of bands in each lane.

4.   Results and Discussion

Variation in the quality of the individual preparations
is normally very evident from the stained gels. Some
preparations will show sharp bands against a clear background, whereas others will appear as a brightly stained
smear with the bands only weakly standing out. Nonetheless, it should generally be possible to observe a pattern
similar to that shown in Fig. 1. Lane 1 corresponds to tube
D; the other three lanes may, of course, be in a different
order.

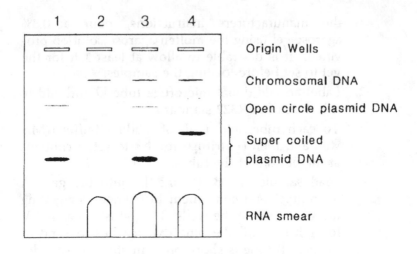

Lane 1   pBR322 standard

Lane 2   JA221 (plasmid free)

Lane 3   JA221 + pBR322

Lane 4   JA221 + yIP5 (pBR322 with yeast DNA insert)

Fig. 1. Result of typical minigel.

Apart from an RNA smear at the distant end of the gel, the most obvious feature should be the supercoiled plasmid bands. Closer to the origin, a generally less sharp chromosomal DNA band should appear. In a good preparation this will be weaker than the plasmid bands.

The three samples should be easy to distinguish. The strain carrying pBR322 should have a band at the same position on the gel as shown by the track containing pure pBR322. The sample carrying the larger plasmid will have a band showing reduced mobility. The absence of any plasmid band should be apparent in the plasmid-free preparation.

Weak bands of open circle plasmid may also be seen closer to the origin (as described in Chapter 3), but these should not obscure the overall picture.

Chapter 10

Isopycnic Centrifugation of DNA

Extraction and Fractionation of Melon Satellite DNA by Buoyant Density Centrifugation in Cesium Chloride Gradients

Donald Grierson

1. Introduction

In order to study the organization and properties of DNA, it is necessary to extract it from cells and purify it from the proteins with which it is usually associated. Cells of plants and animals contain from 1 to 100 pg of DNA per nucleus. This does not sound like a lot until one realizes that is represents 100–10,000 times as much DNA as in an *Escherichia coli* chromosome. When faced with such vast complexity, it is obviously an advantage if the DNA can be separated into a number of more homogeneous fractions, thus simplifying genome analysis. One method of doing this is to exploit the density differences between various gene sequences by centrifuging the DNA in solutions of cesium salts, either alone or in the presence of various heavy metal ions, DNA-binding dyes, or antibiotics that modify density. When solutions of CsCl are centrifuged at sufficient speed, a concentration gradient is formed so that the solution at the bottom of the centrifuge tube is denser than that at the top. Under appropriate conditions, i.e., when the density range in the gradient corresponds to that of DNA, different gene sequences occupy different positions corresponding to their own density. This is called isopycnic or buoyant density centrifugation. It is an equilibrium method: DNA never pellets on the bottom of the tube because the density of the solution is too high. In addition to fractionating the DNA, CsCl centrifugation also separates it from contaminating RNA and carbohydrate, thus enhancing purification.

The buoyant density of DNA is the result of the interaction between the DNA, salt in the medium, and molecules of water. It therefore depends upon the composition of the medium and is different for various cesium salts. For example, the buoyant density of bacteriophage T_4 DNA in CsCl is 1.70 g/mL and in Cs_2SO_4, 1.44 g/mL. The buoyant density is related to the GC content. At neutral pH in CsCl:

$$Density = 0.099 (G + C\%) + 1.660$$

This means that DNAs of different base composition can be conveniently separated from each other. The technique can be extended by including chemicals in the gradient that selectively bind to subfractions of the DNA, thus altering their density. Examples include Ag^+ (used with Cs_2SO_4 gradients) and actinomycin D (used with CsCl gradients). The intercalating dye, ethidium bromide, is also used in CsCl gradients, in which it is especially useful in separating molecules of the same base composition existing in different physical states (i.e., linear molecules, open circles, twisted circles) that affect their dye-binding capacity and thus their density (*see* Chapter 2).

When DNA sequences from a particular organism are centrifuged in CsCl they form a more or less symmetrical peak in the gradient, called the "main band" DNA. Molecules on one side of the peak differ in density from those on the other. DNA from *some* organisms also forms a second "satellite" peak (Fig. 1) and in a few cases several satellites are found. These may be denser or less dense than the main band, depending on their GC content. The definition of a satellite DNA is thus an operational one: It is a subfraction of the DNA that has a different buoyant density from that of the bulk of the DNA. Not all organisms have obvious satellites. A satellite may represent mitochondrial or plastid DNA, a distinct subfraction of

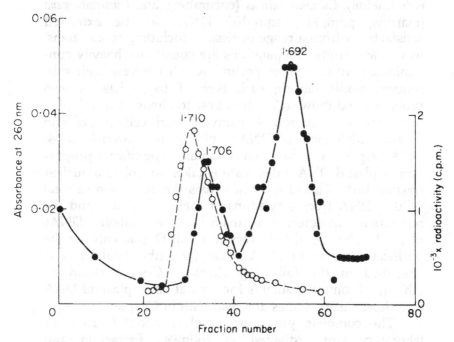

Fig 1. The separation of a mixture of *E. coli* and melon DNA in cesium chloride. A trace of *E. coli* DNA labeled with [³H]thymidine was mixed with melon DNA and centrifuged at 30,000 rpm (57,500g) at 25°C for 80 h in a fixed-angle ultracentrifuge rotor. Three-drop fractions were collected, diluted with 1 cm³ of 50 mM Tris-HCl pH 7.5 containing 1 mM EDTA, and melon DNA located by the absorbance at 260 nm (solid circles, continuous curve). *E. coli* DNA was detected by liquid scintillation counting of aliquots of each sample (open circles, dashed curve). Figures above each peak represent the calculated buoyant density in g/mL (reproduced from ref. 5).

nuclear DNA, or even the genome of a contaminating microorganism.

Centrifugation in CsCl has been used to isolate DNA coding for ribosomal RNA from *Xenopus laevis* (the first eukaryotic gene sequence to be isolated and characterized) (1), for the isolation of animal and plant satellite DNAs (2,3) and to distinguish chloroplast from nuclear DNA (4).

Since plants are inexpensive and easy to grow, they are very suitable sources of DNA for class experiments. Many plants contain satellite DNA, but the highest proportion is found in members of the Cucurbitaceae, such as *Cucumis melo* (melon), *Cucumis sativus* (cucumber), and *Cucurbita pepo* (marrow, pumpkin, squash). DNA can be extracted satisfactorily from a range of tissues, including roots, stems, leaves, and fruits. Preparations are sometimes heavily contaminated with starch or pectins, which interfere with subsequent manipulations; each type of tissue has its own problems and drawbacks. It is best to choose a developing tissue that is composed of relatively small cells in order to obtain a high yield of DNA/g of starting material. Most DNA preparations from plants contain a significant proportion of plastid DNA unless care is taken to isolate a nuclear fraction first. This chapter describes the extraction of total cellular DNA from cotyledons of melon seedlings and the separation and detection of main band and satellite DNAs in CsCl. The use of CsCl–actinomycin D gradients for the purification of plant DNA coding for ribosomal RNA is described in the following chapter. Cesium chloride–ethidium bromide gradients for separation of plasmid DNA from host chromosomes are dealt with in Chapter 2.

The complete procedure involves about 9–10 h of laboratory work, divided as follows. Extraction and purification of the DNA takes about 4 h. After this stage, the DNA may be stored if necessary. Measurement of the UV spectrum, estimation of the amount of DNA recovered, and preparation of the DNA–cesium chloride solutions and centrifuge tubes (so that they are ready for the ultracentrifuge) takes a further 3 h. Ultracentrifugation can begin immediately or be delayed until convenient. After the tubes have been centrifuged for 48 h it takes 2–3 h to fractionate the gradients, locate the DNA, and determine buoyant density.

2. Materials Provided

2.1. Plant Material

Use melon seedlings germinated and grown under aseptic conditions for about 1 wk in the dark.

2.2. Solutions and Chemicals

Homogenizing medium
Phenol - chloroform mixture
Ethanol
80% Ethanol
Ribonuclease buffer
Ribonuclease stock solution
DNA buffer
Cesium chloride

2.3. Apparatus and Equipment

Small mortar and pestle
Glass centrifuge tubes, caps, and racks
Forceps
Pasteur pipets with rubber teats to fit
Ice bucket
Bench centrifuges
Fine balance
Refractometer
Waterbath set at 37°C
Ultracentrifuge, fixed-angle rotor, tubes, and caps
Spectrophotometer and quartz glass cells
Gradient-fractionating equipment

3. Protocol

Do not pipet solutions by mouth. Wear disposable gloves and safety glasses when using the homogenizing medium and phenol–chloroform mixture. If you splash these reagents on your skin, wash thoroughly with water.

The extraction can be performed at room temperature.

1. Harvest the seedlings and collect between 0.4 and 0.6 g of cotyledons after removing the hard seed coat. Homogenize the cotyledons with a mortar and pestle using 1 mL of homogenizing medium for every 0.1 g of tissue. As the DNA and other macromolecules are released into the homogenizing medium, the solution becomes very viscous. Try to disrupt the tissue effectively with a few strokes of the pestle so as not to shear the DNA too much.

2. Pour the slurry into a glass centrifuge tube. Wash the mortar with a small volume of fresh homogenizing medium and add this to the homogenate. Add to the centrifuge tube a volume of phenol–chloroform equal to the volume of the homogenate. Seal the tube with a tight-fitting cap. Do not use parafilm, which is attacked by chloroform. Shake the contents of the tube vigorously for about 5 s to form an emulsion. Centrifuge the tube for 15 min at 2000g.

3. Centrifugation separates the contents of the tube into several zones. The lower liquid phase is phenol–chloroform with dissolved lipids and pigments. At the bottom of the tube is plant debris, including some tissue segments and starch grains. The upper aqueous layer contains water-soluble compounds, including the nucleic acids. Between the aqueous layer and the phenol–chloroform is a white protein precipitate. You may see further plant debris just below this. Gently remove the top aqueous layer with a Pasteur pipet, taking care to leave behind the white protein precipitate. Do not worry if the aqueous layer is cloudy. Transfer it to a clean, glass centrifuge tube, add an equal volume of phenol–chloroform, cap the tube, and shake for about 5 s to form an emulsion. Centrifuge for 15 min at 2000g.

    The aqueous layer should separate clearly from the lower phenol–chloroform layer. There may be a small amount of white protein precipitate at the interface. Carefully transfer the top layer to a clean, glass centrifuge tube with a

Pasteur pipet, leaving the protein behind. Add 2.5 vol ethanol, cap the tube, and *thoroughly* mix the contents by inverting the tube four or five times. You should see a white fibrous precipitate form immediately. This is DNA with some RNA and carbohydrate. Place the tube in an ice bath for 10 min. You should now see a much finer particulate white precipitate forming. This is mainly RNA.

4. Pellet the nucleic acid by centrifugation at 2000g for 10 min. Pour off the liquid and drain the tube upside down for a few minutes. The nucleic acid pellet should remain stuck to the bottom of the tube. Add 10 mL 80% ethanol to the pellet, cap the tube, and shake vigorously to wash the pellet. Traces of phenol, chloroform, and detergents dissolve in the 80% ethanol. Centrifuge the tube at 2000g for 10 min to re-pellet the nucleic acid, pour off the liquid, and drain the tube upside down for a few minutes.

5. Add 3 mL of ribonuclease buffer to the nucleic acid pellet. Shake the tube to dissolve the pellet. This may take a few minutes. Add 0.3 mL of ribonuclease stock solution, gently mix the contents, and place in a water bath at 37°C for 40 min to 1 h.

6. After incubating the solution with ribonuclease, add 3 mL phenol–chloroform solution, cap the tube, and shake vigorously for 5 s to inactivate and precipitate the ribonuclease. Centrifuge the tube for 15 min at 2000g. Carefully remove the top aqueous layer, leaving the white protein precipitate behind, and transfer it to a clean centrifuge tube. Gently layer 7.5 mL ethanol over the top of the aqueous layer. Look at the tube contents carefully. You should see the interface between the two solutions appear slightly cloudy and you may see traces of a white fibrous precipitate. Gently swirl the tube to encourage slow mixing of the solutions at the interface. A white fibrous precipitate will begin to form. Continue swirling the tube, more violently now, until the

contents are mixed. This may take 1 or 2 min. During this time the DNA will form a clump of white fibers. Small fragments of RNA and some carbohydrate may also form a cloudy precipitate if you wait too long.

7. As soon as the precipitate of DNA fibers has formed a clump, remove this with a clean pair of forceps and place it in a fresh glass centrifuge tube containing 10 mL of 80% ethanol. The DNA may be stored indefinitely in a refrigerator until required, or it can be used immediately for the next step.

8. Centrifuge the tube at 2000g for 5 min to pellet the DNA. Pour off the liquid and drain the tube for a few minutes. Add 5 mL DNA buffer, and shake the tube gently to dissolve the DNA. This may take several minutes. Take 0.1 mL of the solution, dilute to 1 mL with DNA buffer, and measure the UV spectrum from 200 to 320 nm using DNA buffer as a blank. You should obtain a typical nucleic acid spectrum with a trough at 230 nm, a peak at 260 nm, and almost zero absorbance at 320 nm. A good preparation will have a 260:230 nm ratio of about 2. If the ratio is much lower than this, it indicates significant carbohydrate contamination, as does absorbance at 320 nm. A shoulder in the spectrum at 280 nm indicates significant protein contamination. An approximate estimate of the amount of DNA recovered can be obtained from the absorbance at 260 nm. Fifty micrograms of DNA dissolved in a 1-mL solution has an approximate absorbance at 260 nm in a 1-cm thick cell of one.

9. Take a volume of DNA solution equivalent to 150–200 μg DNA, pipet into an ultracentrifuge tube, and adjust the final volume accurately to 4 mL with DNA buffer. Weigh out 5.03 g cesium chloride carefully and add this to the DNA solution. Shake the tube gently until the cesium chloride dissolves. The process is endothermic and may be hastened by placing the tube in a

warm waterbath. When the salt has dissolved, take a small drop of solution and measure the angle of refraction with a refractometer at 25°C. This should be approximately 10°0′ (*see* Table 1). If the reading is significantly different from this, add drops of DNA buffer or cesium chloride crystals until the angle of refraction is within the 10°0′–10°8′ range.

10. Layer liquid paraffin over the solution in the ultracentrifuge tube and fit and tighten the tube cap. Fill the tube with liquid paraffin to the top and fit the grub screw in the tube cap. Clean the outside of the tube and cap. Weigh the tube and check that the weight is within 0.1 g of the tube to be placed opposite it in the ultracentrifuge rotor (*see* Appendix I, "Using an Ultracentrifuge").

   The samples can be stored, or placed in the ultracentrifuge immediately. Use a fixed-angle ultracentrifuge rotor and spin at about 30,000 rpm (57,000–60,000g) for 2 d at 25°C.

11. At the end of the ultracentrifuge run, fractionate the gradients quickly. Do not store the tubes for long periods of time. There are several methods for fractionating and analyzing gradients, involv-

Table 1
Relationship Between CsCl Density, Refractive Index,
and Degrees and Minutes at 25°C[a]

Density[b], g/mL	Refractive index, 25°C	Degrees and minutes
1.4000	1.3722	7° 16′
1.5000	1.3815	8° 12′
1.6000	1.3905	9° 8′
1.7000	1.3996	10° 4′
1.8000	1.4086	11° 0′

[a] Intermediate figures can be calculated by interpolation from a graph.

[b] Density = (10.8601 × refractive index) − 13.4974.

96 Grierson

ing the use of flow-cells, UV analyzers, and fraction collectors. A simple manual method is described here (Fig. 2). Clamp the ultracentrifuge tube in a vertical position and remove the grub-screw from the cap. Gently lower a long hollow metal needle through the hole in the cap until it comes to rest on the bottom of the tube. Clamp the needle in position and attach it to a peristaltic pump with microbore tubing. Switch on the pump and collect 5-drop fractions in small test tubes or vials. You will need about 60 numbered tubes in a small rack. Save some of the samples for measurement of density. Cap every tenth tube immediately to prevent evaporation of solution and later measure the angle of refraction of the solution with a refractometer. Dilute the remaining samples with 1 mL distilled water and measure the absorbance at 260 nm.

Save each sample after measuring the absorbance: the fractions can be used for studying DNA dissociation–reassociation, DNA–RNA hybridization, or restriction endonuclease digestion and cloning.

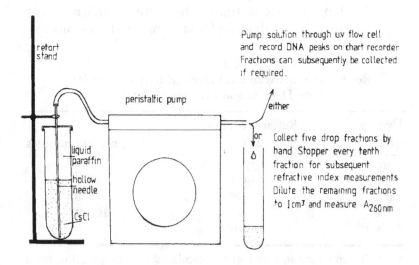

Fig. 2. A simple method for fractionating cesium chloride gradients.

4. Results and Discussion

Plot a graph of the absorbance of each gradient fraction at 260 nm against fraction number. A typical result, with a single peak of melon satellite DNA (density 1.706 g/mL) well separated from the main band DNA (density 1.692 g/mL), is shown in Fig. 1. In this experiment a trace of radioactive *E. coli* DNA was also included in the gradient.

High absorbance toward the bottom (left in Fig. 1) of the gradient is caused by contaminating RNA that was not removed by ribonuclease. RNA is denser than DNA in cesium chloride and pellets on the bottom of the centrifuge tube. Incomplete digestion products of RNA are often found near the bottom of the gradient. If high absorbance is observed near the top of the gradient, this is probably caused by contaminating carbohydrate.

Melon DNA should resolve into two peaks, as shown in Fig. 1. Cucumber DNA, on the other hand, has two satellite DNAs with density 1.701 and 1.707 g/mL (not shown). If you find an extra peak of absorbance between the satellite DNA and the bottom of the gradient, this is almost certainly caused by DNA from bacteria that were growing on the melon seedlings.

If undiluted samples were used for measurement of their angle of refraction with a refractometer, the density at intervals down the gradient can be calculated. It is important that evaporation from the solution was prevented before measurements were taken. From the information in Table 1, convert the refractometer readings (in degrees and minutes) to density. Plot the density values corresponding to the appropriate fractions on your graph. By interpolation, calculate the density of the main band and satellite DNAs.

This result shows that melon DNA can be fractionated into two distinct components that differ in buoyant density and, therefore, GC content. Which is the most likely: that the two types of sequences are closely interspersed in the chromosomes or that sequences of a particular GC content are clustered together, and are separate from each other? What additional information would you require in order to help to distinguish between these two possibilities?

Melon satellite DNA has been studied by several groups (6,7). It consists of repeated DNA sequences of two different types. It has been suggested, at least in cucumber, that the satellite DNAs may originate from chloroplasts and/or mitochondria (8).

References

1. Birnstiel, M., Speirs, J., Purdom, I., Jones, K., and Loening, U. E., (1968) Properties and composition of the isolated ribosomal DNA satellite of *Xenopus laevis*. *Nature* **219**, 454–463.

2. Flamm, W. G. (1972) Highly repetitive sequences of DNA in chromosomes. *Int. Rev. Cytol.* **32**, 2–51.

3. Ingle, J., Pearson, G. G., and Sinclair, J. (1973) Species distribution and properties of nuclear satellite DNA in higher plants. *Nature New Biol.* **242**, 193–197.

4. Wells, R. and Ingle, J. (1970) The constancy of the buoyant density of chloroplast and mitochondrial deoxyribonucleic acid in a range of higher plants. *Plant Physiol.* **46**, 178–179.

5. Grierson, D. (1977) The Nucleus and the Organization and Transcription of Nuclear DNA, in *The Molecular Biology of Plant Cells*. (H. Smith, ed.) Blackwell, Oxford.

6. Sinclair, J., Wells, R., Deumling, B., and Ingle, J. (1975) The complexity of satellite deoxyribonucleic acid in a higher plant. *Biochem. J.* **149**, 31–38.

7. Bendich, A. J. and Taylor, W. C. (1977) Sequence arrangement in satellite DNA from the musk melon. *Plant Physiol.* **59**, 604–609.

8. Kadouri, A., Atsmon, D., and Edelman, M. (1975) Satellite-rich DNA in cucumber: Hormonal enhancement of synthesis and subcellular identification. *Proc. Natl. Acad. Sci. USA* **72**, 2260–2264.

Chapter 11

Purification of Plant Ribosomal DNA

Use of Buoyant Density Centrifugation in Cesium Chloride–Actinomycin D Gradients and DNA–RNA Hybridization

Donald Grierson

1. Introduction

The general features of isopycnic centrifugation of DNA and the separation of main-band from satellite DNAs in solutions of cesium chloride are discussed in the preceeding chapter. This chapter describes a modification of the procedure that enhances the separation of specific DNA sequences from main-band DNA. The method is based on the fact that actinomycin D binds preferentially to DNA molecules with a high-GC content. This *lowers* their density relative to main-band and other satellite DNAs so that they can be resolved separately. Cesium chloride-actinomycin D gradients have been used to isolate sea urchin histone genes (1) and plant ribosomal DNA (2), and to separate genes coding for plant 25S and 18S from 5S RNA genes (3) and as a prelude to the cloning of ribosomal DNA (4).

The method for extraction and purification of melon DNA is exactly as described in Chapter 10, but the preparation of the DNA–cesium chloride solution for buoyant density centrifugation is different. The purification procedure works so effectively that it is difficult to detect the separated rRNA genes by their UV absorbance because they are present only in small quantities. For this reason a DNA–RNA hybridization procedure for the detection of the rRNA genes is described.

A total of 12 h of laboratory work, separated by a 72-h centrifugation step, is required to complete this experiment, excluding the preparation of radioactive rRNA. The procedure can be conveniently accomplished in several sessions lasting 2–4 h.

2. Materials Provided

2.1 DNA Isolation and Buoyant Density Centrifugation

All items listed in section 2, Chapter 10.

2.2. DNA-RNA Hybridization

The following additional items are provided:

5M Sodium hydroxide
Neutralizing solution
Nitrocellulose filters
Swinnex filter holders and 10- or 20-mL disposable syringes
2× Standard saline citrate (2 × SSC)
Water baths at 70 and 37°C
Radioactive rRNA
Scintillation vials
A toluene-based scintillation fluid and vials
Desiccators
Vacuum oven

3. Protocol

For extraction and purification of DNA, follow the procedure outlined in section 3 of Chapter 10 to the end of step 7, and continue as follows.

1. Centrifuge the DNA at 2000g for 5 min, pour off the liquid, and drain the tube for a few minutes. Dissolve the DNA in 6 mL DNA buffer. It may be necessary to shake the tube for several minutes before the DNA dissolves. Dilute a 0.1-mL aliquot with 1 mL buffer, and measure the UV spectrum from 230 to 320 nm. Calculate the DNA concentration (50 μg DNA dissolved in 1 mL has an absorbance of 1 (approximately) at 260 nm in a 1-cm light path).

2. Pipet 4.0 mL of DNA solution, containing 100–200 μg DNA, into an ultracentrifuge tube, diluting the solution with DNA buffer if necessary. Weight 4.0 g cesium chloride, add to the DNA solution, and shake gently to dissolve. Cool the solution in an ice bath and slowly add 0.125 mL actinomycin D solution (stock solution 1 mg/mL). Measure the angle of refraction of a small drop of the solution at 5°C using a refractometer, and convert this to a refractive index value using the refractometer tables (Chapter 10, Table 1). The refractive index should be 1.3900 at 5°C. Add DNA buffer or cesium chloride if the measured value is significantly different from this.

3. Layer liquid paraffin over the DNA solution and fit and tighten the tube cap. Fill to the top with liquid paraffin, excluding all air bubbles from the tube. Fit the grub screw in the cap and clean the outside of the tube and cap. Weigh the tube and check that it is within 0.1 g of the weight of the tube to be placed opposite in the ultracentrifuge rotor (see Appendix I, "Using an Ultracentrifuge"). Keep the tubes cold. The samples can either be stored like this for some days or placed in the ultracentrifuge immediately. Centrifuge the tubes

in a fixed-angle rotor at 3–4°C for 3 d at 30,000 rpm (57,000–60,000g).

4. Fractionate the gradients as soon as possible after the ultracentrifuge rotor stops. Collect five-drop fractions from the gradients, as described in section 3, step 11 of Chapter 10. Dilute the fractions with 1 mL DNA buffer and measure the absorbance at 260 nm.

5. The DNA sequences coding for rRNA are detected by hybridization with radioactive rRNA. The result in Fig. 1 was obtained by hybridizing a sample of *each* gradient fraction to rRNA. However, preparation of the samples is time consuming and the method can be simplified by hybridizing only a small number of pooled fractions. First, plot a graph of the absorbance at 260 nm of each gradient fraction to see where the DNA peaks are. Pool fractions corresponding to 1–20, 21–35, 36–50, 55–75 in Fig. 1D, measure the volume and UV absorbance, and calculate the total amount of DNA in each fraction.

6. For hybridization, the DNA has to be rendered single-stranded by treatment with alkali and attached to nitrocellulose filters. Take one nitrocellulose filter for each DNA fraction you have and make an identification mark on the edge of each one with a pencil. Take care not to crack the filters. Place each filter carefully in a swinnex filter holder, using forceps, add the sealing ring, and gently screw the two halves of the holder together. Attach a 10 or 20 mL disposable syringe to each filter holder after removing the plunger, and stand the filter holder/syringe assemblies vertically in a suitable rack. Add 2 mL 2 × SSC to each syringe and allow to drain through the filters.

7. To each mL of DNA solution, add 0.2 mL 5M sodium hydroxide and mix thoroughly (measure the volume of sodium hydroxide carefully). The high pH dissociates double-stranded DNA. Cool the solution in an ice bath, allow to stand for 15 min, then bring to pH 8 by adding 4 mL cold

neutralizing solution for every 0.2 mL 5M sodium hydroxide added previously. Keep the solution cold to reduce the DNA renaturation rate. As soon as possible, pour the neutralized DNA solutions, in batches if necessary, into the syringes and allow to drain through the nitrocellulose filters. As this happens, the single-strand DNA sticks to the filters. If the solution takes too long to run through, push it through *very slowly and gently* using the syringe plunger. When all of the solution has drained through the filter, unscrew the filter holder and remove the filter with a pair of forceps. Lay the filters, DNA up, on tissue paper in a desiccator and dry under vacuum for 30 min, then transfer to a vacuum oven and heat at 80°C for 2 h or more. The filters can be stored dry in a refrigerator until required for hybridization.

8. For each hybridization reaction, you need one scintillation vial containing 3 mL of a 5 μg/mL solution of radioactive rRNA dissolved in 2 $\times$ SSC. *Wear disposable gloves and take appropriate precautions when working with radioisotopes (see* Appendix I, "Safe Handling of Radioactivity"). Place the DNA filters in the rRNA solution, including one blank filter, seal the vial, and incubate at 70°C in a water bath for 2 h. Remove the filters with forceps and place them together in 500 mL 2 $\times$ SSC. Stir with a magnetic stirrer for 15 min, then transfer the filters to 500 mL fresh 2 $\times$ SSC and repeat the washing. Transfer the filters to a small flask or beaker and add 10 mL 2 $\times$ SSC. Add sufficient ribonuclease from the stock solution to bring the final concentration of enzyme to 10 μg/ mL. Incubate at 37°C for half an hour, with occasional shaking. Remove the filters with forceps and wash once for 15 min in 500 mL 2 $\times$ SSC and once in 1 $\times$ SSC for 15 min. Dry the filters at 40°C in an incubator or under an infrared lamp, and place them in scintillation vials with a toluene-based scintillant. Measure the radioactivity attached to the filters with a liquid scintillation counter.

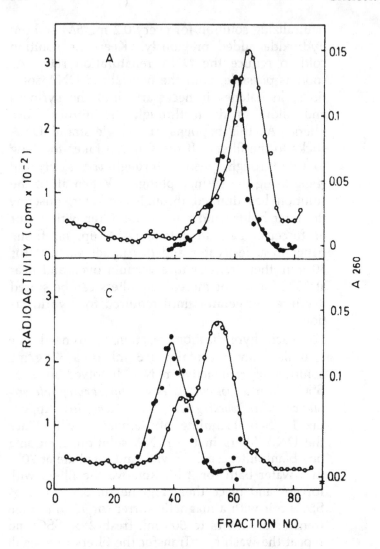

Fig. 1. Separation of rRNA genes from main-band and satellite DNA in cesium chloride and actinomycin–cesium chloride. Unlabeled total DNA from *Matthiola incana* (A,B) and *Cucumis melo* (C,D) was fractionated in cesium chloride (A,C) or actinomycin–cesium chloride (B,D). Centrifugation conditions were 32,000 rpm (60,000g) at 20°C for 60 h (A,B) and 30,000 rpm at 3°C for 85 h (C,D). Seven-drop fractions were collected, diluted with 1 mL DNA buffer, and DNA located by absorbance at 260 nm (open circles). Each fraction was saved and the DNA denatured, attached to nitrocellulose filters, and hybridized to (^{3}H)-labeled rRNA. Hybridized RNA was detected by scintillation counting (solid circles) (reproduced from ref. 2).

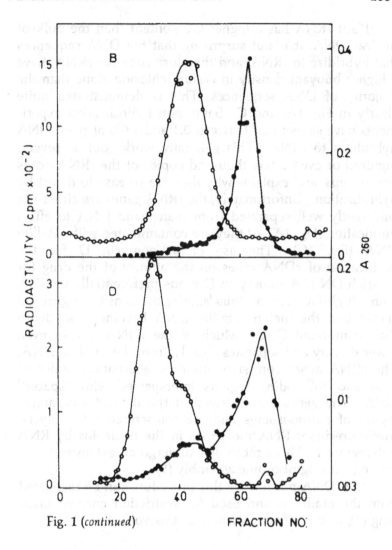

Fig. 1 (continued)   FRACTION NO.

4. Results and Discussion

Plot a graph of the absorbance at 260 nm of each gradient fraction. Indicate the extent of hybridization of each set of pooled fractions with radioactive rRNA as a histogram. To obtain a measure of the purity of the rRNA genes in each fraction, plot the cpm for rRNA hybridized per μg of DNA. Compare your results with those in Fig. 1B and C.

Plant rRNA has a higher GC content than the bulk of nuclear DNA. It is not surprising that the DNA sequences that hybridize to rRNA, and therefore code for rRNA, have a higher buoyant density in cesium chloride alone than the majority of DNA sequences. This is demonstrated quite clearly in Fig. 1A, and C. Saturation hybridization experiments have shown that between 0.1 and 1.0% of plant DNA hybridizes to rRNA. This generally works out as several hundred or even a few thousand copies of the rRNA genes per nucleus, and explains why they are so easy to detect by hybridization. Unfortunately, the rRNA genes are either not sufficiently well separated from main-band DNA to effect purification (Fig. 1A), or they are contaminated with satellite DNA (Fig. 1C). The use of actinomycin D for the purification of rDNA relies on the affinity of the drug for GC-rich DNA. Actinomycin D binds preferentially to DNA with a high-GC content, thus *lowering* the density. Figure 1D shows that the melon satellite DNA becomes less dense than main-band DNA. Much of the rDNA has an even lower density and separates cleanly from the satellite DNA. The rRNA genes consist of multiple alternating copies of 25S- and 18S-coding regions interspersed with "spacer" DNA. The genes are clustered at the nucleolar organizer region of chromosomes and are transcribed into polycistronic precursor RNA molecules in the nucleolus by RNA polymerase I. These molecules undergo an extensive "processing" during ribosome assembly (5).

The rDNA purified by this procedure can be recovered from the gradients and used for restriction enzyme mapping (4), as described elsewhere in this volume.

References

1. Birnstiel, M., Telford, J., Weinberg, E., and Stafford, D. (1974) Isolation and some properties of the genes coding for histone proteins. *Proc. Natl. Acad. Sci. USA* **71**, 2900–2904.
2. Hemleben, J., Grierson, D., and Dertmann, H. (1977) The use of equilibrium centrifugation in actinomycin–cesium chloride for the purification of ribosomal DNA. *Plant Sci. Lett.* **9**, 129–135.

3. Hemleben, V. and Grierson, D. (1978) Evidence that in higher plants the 25S and 18S rRNA genes are not interspersed with genes for 5S rRNA. *Chromosoma (Berlin)* **65**, 353–358.

4. Friedrich, J., Hemleben, V., Meagher, R.G., and Key, J.L. (1979) Purification and restriction endonuclease mapping of soybean 18S and 25S ribosomal RNA genes. *Planta* **146**, 467–473.

5. Grierson, D. (1982) RNA Processing and Other Posttranscriptional Events, in: *Nucleic Acids and Proteins in Plants*, Vol. II (Parthier, B. and Boulter, D., eds.) Springer-Verlag, Berlin-Heidelberg.

Chapter 12

The Extraction and Affinity Chromatography of Messenger RNA

Robert J. Slater

1. Introduction

Living cells contain two forms of nucleic acid—deoxyribonucleic acid (DNA) and ribonucleic acid (RNA). It has been known for a number of years that RNA molecules are involved in the expression of the genetic material (DNA) and are, in fact, the initial products of gene expression that go on to direct protein synthesis. Knowledge of the structure, metabolism, and mode of action of RNA in protein synthesis is therefore fundamental to our understanding of the control of gene expression, cell division, growth, and development. Furthermore, the recent spectacular advances in genetic engineering depend on a working knowledge of RNA synthesis and its control.

There are many different types of RNA present in the cell, and they are generally classified according to function or size, i.e., molecular weight or sedimentation coefficient (S). The three principal types of RNA, according to function, are ribosomal, transfer, and messenger RNA (rRNA, tRNA, and mRNA, respectively). Ribosomal RNA molecules are structural components of ribosomes and constitute the bulk (approximately 75%) of cellular RNA. They are present in the ribosomes as three different molecules, classified by size: 28S (or 25S), 18S, and 5S in eukaryotic cytoplasmic ribosomes, or 23S, 16S, and 5S in prokaryotic-type ribosomes. Transfer RNAs are smaller molecules with a molecular weight of approximately 25,000, and are responsible for the supply of specific amino acids to the ribosomes. Messenger RNA is generally present as a small percentage (1–5%) of the total cellular nucleic acid and consists of a heterogeneous mixture of molecules that code for the different proteins synthesized on the ribosomes. The mRNA fraction of cells thus consists of thousands of molecules, differing in base composition and size, and cannot therefore be fractionated as a single band on polyacrylamide or agarose gels (*see* Chapter 13). Messenger RNA molecules of eukaryotic organisms do, however, have a number of features in common. They have a m^7Gppp cap (7-methylguanosine joined to the mRNA by a triphosphate link following transcription) on the 5′ end, which is thought to increase the stability of the molecules by protecting them from phosphatases and nucleases, and the vast majority have a poly(A) tail at the 3′ end. This poly(A) sequence of up to 250 bases is added, following transcription, by poly(A) polymerase in the nucleus. Experiments involving the injection of mRNA into *Xenopus laevis* oocytes have shown that the poly(A) tail enhances the stability, but is not essential for the translation of mRNA (1). Histone mRNAs, for example, do not have a poly(A) tail. The presence of poly(A) tracts in mRNA molecules forms the basis of a quick and convenient affinity chromatography method for the separation of mRNA from other forms of nucleic acid (2).

The ability to isolate mRNA is an important aspect of molecular biology research. For example, characterization of mRNA preparations in an in vitro protein-synthesizing system is a common approach to the study of changes in gene expression that occur during growth and development or in response to hormones. Alternatively, the mRNA can be used as a template for the synthesis of complementary DNA molecules. If these cDNAs are prepared from a total mRNA preparation, they can be cloned to form a cDNA "library" that forms a source of individual cDNA clones of interest. In some cases, for example in the study of hemoglobin gene expression, specific mRNA has been isolated that codes for a single protein. A cDNA made from the mRNA can be cloned and, following radiolabeling, used as a hybridization probe to locate the genomic DNA sequence. A detailed description of these techniques is beyond the scope of this chapter. For a more detailed discussion of in vitro translation, the use of cDNA libraries and the isolation of specific mRNAs, refer to refs. 3, 4, and 5, respectively.

In general, methods for the extraction of RNA involve the treatment of cell homogenates with a deproteinizing medium that digests or denatures protein and leaves the RNA in solution (6). Deproteinizing agents can be enzymes, such as proteinase K, or chemical protein denaturants, such as phenol. The experiment described here relies on the homogenization of plant tissue in a cell-lysing solution that contains a chelating agent (sodium aminosalycilate) and a strong detergent (sodium tri-isopropylnaphthalene sulfonate—TPNS or sodium dodecyl sulfate—SDS) to break the cells, disrupt nucleo–protein complexes, and inhibit ribonuclease activity. A water-saturated phenol mixture is then added to denature protein and extract hydrophobic components. Following centrifugation, organic and aqueous phases separate and DNA and RNA can be recovered by precipitation from the upper, aqueous phase by the addition of absolute alcohol, thereby separating the nucleic acids from small molecular weight contaminants such as carbohydrates, amino acids, and nucleotides.

The mRNA molecules that contain poly(A) sequences can be purified by passing a solution of the extracted nucleic acids through a column containing oligo(dT)-cellulose (2). Under high-salt conditions (0.3–0.5M NaCl or KCl), which favor nucleic acid hybridization, poly(A)-containing RNA will bind to the column, but DNA and the rest of the RNA will pass straight through. Following thorough washing of the column, mRNA is recovered by eluting with a low-salt buffer. Poly(A)-containing RNA prepared in this way can be used as a template for in vitro protein synthesis or for cDNA preparation.

The experiment described here involves the extraction of total nucleic acids from embryo axes of germinating pea seeds followed by affinity chromatography of the poly(A)-containing RNA. The mRNA content of most cells is quite low, so the technique is best illustrated using radiolabeled material. If preferred, however, the experiment can be performed with unlabeled material, but a very sensitive UV spectrophotometer will be required to monitor the nucleic acid fractions isolated. The experiment is carried out on two separate occasions that can be up to 14 d apart if ^{32}P is used, or longer if not. The procedure for the first day involves the extraction of total nucleic acids and should take no longer than 2 h. The second day involves oligo(dT)-cellulose chromatography to separate the poly-(A)-containing RNA from the bulk of the nucleic acids, and will take up to 3 h. If desired, the prepared nucleic acids can be analyzed by gel electrophoresis, combining this experiment with that described in Chapter 13. This extra procedure is most conveniently done during a third session of 3–4 h.

2. Materials Provided

2.1. Day 1

    Sterile germinated pea seeds incubated for 4 h at 17–20°C in sterile distilled water containing 50 µCi/mL ^{32}P-orthophosphate

Homogenizing solution
Deproteinizing solution
Absolute ethanol
Pestle and mortar
Glass centrifuge tubes and caps
Petri dish, scalpel, and forceps
Rubber gloves and safety glasses

2.2. Day 2

70% Ethanol containing 0.5% SDS
Oligo(dT)-binding buffer
Oligo(dT)-elution buffer

Small glass chromatography column and stand
20 Small test tubes
Oligo(dT)-cellulose, swollen in elution buffer
Glass fiber filters and filter apparatus
Bovine serum albumin solution (500 μg/mL)
Ice-cold 10% TCA
Ice-cold 5% TCA
Absolute ethanol
Scintillation vials and caps
Scintillation fluid
Rubber gloves and safety glasses

3. Protocol

3.1. Day 1: The Extraction of Total Nucleic Acids

The detergent and deproteinizing solutions contain phenol: gloves and safety glasses must be worn. The plant material has been incubated in ^{32}P-orthophosphate. Take note of the safety precautions pertinent to the use of radioisotopes described in Appendix I.

1. Wearing rubber gloves, select three or four pea seedlings with a pair of forceps and place them in a Petri dish. Remove the radicles using the scalpel and forceps, and transfer to the mortar. Discard the Petri dish and cotyledons in the waste receptacle provided.

2. Grind the radicles in 1–2 mL homogenizing solution using the mortar and pestle. Successful lysis of the cells is accompanied by an increase in the viscosity of the solution caused by the release of DNA and other macromolecules.

3. Transfer the homogenate to a glass centrifuge tube. Rinse out the mortar with a small volume of homogenizing solution and add this to the homogenate. Place the contaminated mortar and pestle in the detergent washing-up solution provided.

4. Add to the centrifuge tube a volume of deproteinizing solution equal to the volume of the homogenate and seal the tube with a cap (do not use sealing films, which dissolve in chloroform). Shake the tube for 5 min to maintain an emulsion.

5. Remove the cap with the aid of a paper tissue, to absorb any potential aerosol, and spin the tubes for 10–15 min in a bench centrifuge at 2000g or top speed. This separates the tube contents into three phases, aqueous, denatured protein, and phenol, as shown in Fig. 1.

6. Carefully remove the upper, aqueous layer with a Pasteur pipet and retain in a second centrifuge tube containing an equal volume of deproteinizing solution. Do not worry if the aqueous layer is slightly cloudy.

7. Extract the residual nucleic acids from the remaining phenol and protein phases by adding an additional 1 or 2 mL of detergent solution to the original centrifuge tube. Shake, centrifuge, and remove the aqueous phase and add to the second tube as before. This step is important if a quantitative recovery of nucleic acids or total poly(A)-containing RNA is required; otherwise it may be omitted.

8. Re-extract the combined aqueous phases by shaking with the deproteinizing mixture for a further 5

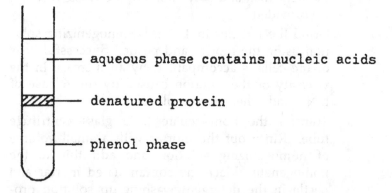

Fig. 1. The appearance of a deproteinized homogenate following centrifugation.

min. Spin the tube in a bench centrifuge and
then, with a Pasteur pipet, carefully transfer the
aqueous phase to a clean centrifuge tube.

9. Add 2.5 vol of absolute alcohol to the aqueous
solution, mix thoroughly, and leave at $-20°C$
overnight to allow precipitation of nucleic acids.
A DNA precipitate resembling cotton wool often
appears immediately upon addition of alcohol, but
the RNA precipitate, resembling snowflakes, takes
several hours to form at $-20°C$. Precipitation is
more rapid at lower temperatures.

10. Dispose of all radioactive waste in the receptacles
provided and soak contaminated glassware and
other apparatus in detergent solution. Monitor
your working area, hands, and clothing with the
radiation detector and wash your hands before
leaving the laboratory.

3.2. Day 2: To Determine the Concentration and Specific Radioactivity of the Nucleic Acid Preparation

1. Collect the nucleic acid precipitate by centrifugation for 10 min at top speed in a bench centrifuge.
Discard the supernatant in the waste receptacle
provided.

2. Wash the precipitate by suspending in 70%
alcohol containing 0.5% SDS, followed by centrifugation as before. This step is designed to
remove phenol and small molecular weight contaminants present in the preparation. Discard the
supernatant in the waste receptacle provided and
support the tube, upside down, over a paper tissue
for a few minutes to drain the precipitate of
alcohol.

3. Dissolve the precipitate in 2 mL of oligo(dT)-
binding buffer. Estimate the concentration of
nucleic acids in solution by diluting a 50-μL aliquot of your solution to 3 mL with distilled water
and measuring the absorbance at 260 and 280 nm.

A pure nucleic acid solution should give a 260:280 nm ratio of 2. A ratio significantly lower than this indicates protein contamination. Assume that an absorbance reading at 260 nm of 1.0 is equivalent to 45 μg/mL.

4. The specific radioactivity of the sample is determined in duplicate as follows. Transfer two 50-μL aliquots to a pair of test tubes. Add 100 μL of bovine serum albumin solution and 1 mL ice-cold 10% TCA to each tube; mix and leave on ice for approximately 30 min. Meanwhile, the oligo-(dT)-cellulose column can be set up as described below, in section 3.3. Collect the precipitates on 2.5-cm glass fiber disks, pre-wetted with 5% TCA, using the apparatus provided. Wash the precipitates with 3 $\times$ approximately 5 mL ice-cold 5% TCA, and finally, 1 $\times$ approximately 2 mL of ethanol. When dry, estimate the radioactivity on the filter by scintillation counting.

5. If the quality of the nucleic acid preparation is to be monitored by gel electrophoresis, transfer two aliquots of the nucleic acid solution containing 2 and 5 μg (calculated in section 3.2, step 3), to small polypropylene tubes. Add a minimum of 2.5 vol. of ethanol to each tube, mix thoroughly, and leave at $-20°C$ overnight. Gel electrophoresis is then carried out according to the procedure described in Chapter 13.

3.3. Day 2: Oligo(dT)-Cellulose Chromatography

1. Place the chromatography column in the stand provided. With the outlet clip closed, pack the column with 25 mg oligo(dT)-cellulose from the slurry provided using a Pasteur pipet. Equilibrate the column by washing through several vol of the oligo(dT)-binding buffer.

2. Load the sample onto the column slowly (approximately 20 mL/h) and immediately begin collecting 1.5-mL fractions. The column effluent can be re-

applied to the column, if desired, to ensure maximum binding of poly(A)-containing RNA.

3. Wash the column with oligo(dT)-binding buffer (the flow rate can now be increased to a convenient level), until 10 1.5-mL fractions have been collected, i.e., when nucleic acids can no longer be detected in the effluent.

4. Elute the column with oligo(dT)-elution buffer until six 1.5-mL fractions have been collected. Stop the column, leaving a layer of elution buffer above the oligo(dT)-cellulose.

5. If possible, determine the A_{260} for the six eluent fractions in a UV spectrophotometer. Pairs of fractions may have to be pooled if only 3-mL cells are available. Do not be surprised if the A_{260} reading is very low.

6. Transfer 1-mL aliquots from each of the 16 fractions into scintillation vials and add the recommended volume of scintillation fluid. Determine the radioactivity in a scintillation counter.

7. Dispose of all radioactive waste in the receptacles provided and soak all contaminated glassware (except the column, which can be used again) in detergent solution. Monitor your working area, hands, and clothing with the radiation detector and wash your hands before leaving the laboratory.

4. Results and Discussion

From the absorbance readings at 260 nm, calculate the total nucleic acid content of the original preparation and the "bound" fractions eluted from the column.

What proportion of the total nucleic acids isolated is poly(A)-containing RNA?

To illustrate the nucleic acid fractionation achieved by the oligo(dT)-cellulose column, use the data obtained from the scintillation counter to plot a graph of dpm/fraction against fraction number. Corrections will have to be made to the data to take account of the fraction volume, counting efficiency, and background count, although the latter two may have been done for you by the instrument automatically. Mark, with an arrow on the graph, the position representing the switch from oligo(dT)-binding buffer to elution buffer. There should be a clear peak of radioactivity immediately after this position.

Use the counts obtained for the glass fiber filters to calculate the total incorporation of radioactivity into the nucleic acids isolated (this eliminates the radioactivity caused by unincorporated ^{32}P). Calculate the specific activity (dpm/μg) of the preparation.

From the data used to plot the graph, calculate the radioactivity incorporated into poly(A)-containing RNA. What proportion of the total radioactivity in the preparation (calculated from the counts obtained on the glass fiber filter) is present in poly(A)-containing RNA? Is this value the same as that calculated using the absorbance data? If not, suggest an explanation. Would you expect the proportion of radioactivity present in the poly(A)-containing RNA to be the same if the peas had been incubated in ^{32}P for 24 h instead of 4 h?

5. Notes

The experiment described here illustrates the use of oligo(dT)-cellulose chromatography to isolate mRNA. The procedure described is essentially the same as that routinely used in many research laboratories. Modifications include an optional heat treatment of the RNA, prior to chromatography, to disrupt aggregates, and rechromatography of the poly(A)-containing RNA to thoroughly remove any contaminating rRNA.

The mRNA fraction isolated can be used for cDNA preparation or in vitro translation. It is essential, however, to remove SDS, included to inhibit RNase activity, prior to carrying out these reactions. RNA is recovered from the low-salt fractions by adding sodium acetate to $0.15M$ and 2.5 vol. of alcohol. Following precipitation at $-20°C$ overnight, the precipitate is washed several times with 70%

alcohol to remove SDS. The mRNA can then be stored as a precipitate under alcohol at $-20°C$ (or lower), or drained and dissolved in sterile distilled water for experimental use.

References

1. Littauer, U.Z. and Soreq, H. (1982) The Regulatory Function of Poly(A) and Adjacent 3′ Sequences in Translated RNA, in *Progress in Nucleic Acid Research and Molecular Biology* Vol. 27 (Coln, W.E., ed.) Academic Press, London and New York.
2. Aviv, H. and Leder, P. (1972) Purification of biologically active globin messenger RNA by chromatography on oligothymidylic acid cellulose. *Proc. Natl. Acad. Sci. USA* **69**, 1408–1412.
3. Grierson, D. and Speirs, J. (1983) Protein Synthesis In Vitro, in *Techniques in Molecular Biology.* (Walker, J.M. and Gaastra, W., eds.). Croom Helm, London and Canberra, and Macmillan, New York.
4. Williams, J.S. (1981) The Preparation and Screening of a cDNA Clone Bank, in *Genetic Engineering* (Williamson, R., ed.). Vol 1., Academic Press, New York.
5. Taylor, J.M. (1979) The isolation of eukaryotic messenger RNA. *Ann. Rev. Biochem.* **48**, 681–717.
6. Slater, R.J. (1983) The Extraction and Fractionation of RNA, in *Techniques in Molecular Biology.* (Walker, J.M. and Gaastra, W., Eds.). Croom Helm, London and Canberra, and Macmillan, New York.

Chapter 13

Agarose Gel Electrophoresis of RNA

Robert J. Slater

1. Introduction

Gel electrophoresis is one of the most important techniques currently available for the fractionation of RNA. The experimental procedure is relatively simple, but nevertheless achieves very reproducible results and high resolution. RNA is a polyanion and will therefore migrate toward the positive electrode in an electric field. If the migration occurs through a gel matrix of carefully chosen pore size, the mobility of the RNA molecules is related to the logarithm of the molecular weights: the smallest molecules moving the greatest distance. Suitable gel matrices for the electrophoresis of RNA are polyacrylamide or agarose in the form of rods or slabs. Agarose is generally preferred to acrylamide because of its ease of handling and lower toxicity, although acrylamide gives better resolution of small molecular weight RNA. Slabs have a number of advantages over rod gels: they enable many samples to be fractionated under identical conditions, they can be easily stained and photographed, and following electrophoresis the nucleic acids can be transferred to cellulose nitrate paper (Northern blotting) for hybridization experiments by a procedure similar to that described in Chapter 5 (*1*). In the procedure described here, total nucleic acids isolated from plant cells will be fractionated in a horizontal agarose electrophoresis system using the apparatus already described in Chapter 3.

A total nucleic preparation from plants will contain the following components: DNA, rRNA, mRNA, and tRNA. In this experiment the DNA will be of high molecular weight and during electrophoresis will migrate only a short distance as a single band. The RNA will fractionate into its component parts according to molecular weight and, following staining with ethidium bromide, will be visible as discrete bands. The mRNA component, however, will not be visible since it is present as only a small proportion (1–5%) of total RNA and is made up of thousands of different molecules. No individual mRNA is therefore present in sufficient quantity to show up as a visible band on staining. Transfer RNAs are all of very similar molecular weight and will run as a single band during electrophoresis. The rRNA, however, is made up of a number of components of differing molecular weight and will therefore separate into several bands.

Ribosomal RNA makes up the bulk (approximately 75%) of cellular RNA and consists of three types of mature molecule classified by size. Eukaryotic cytoplasmic ribosomes contain one molecule each of a 28S (or 25S in plants), 18S, and 5S molecule, whereas prokaryotic (and eukaryotic organelle) ribosomes contain the same ratio of 23S, 16S, and 5S molecules. The two largest molecules in each group arise from the processing of a common, large molecular weight precursor (approximately 45S) that also codes for a small 5.8S rRNA that remains complexed with the 28S (or 25S) molecule through hydrogen bonds following processing. The 5S component is transcribed from a separate gene. These various rRNA species should be visible following electrophoresis and staining.

Gel electrophoresis of RNA can be carried out under nondenaturing or denaturing conditions. In the former case, secondary structure of the nucleic acids is maintained, but under denaturing conditions, hydrogen bonding is disrupted and RNA runs as single-stranded molecules. Under the latter system, therefore, additional bands may be present, such as 5.8S rRNA and breakdown products from rRNA. The choice of running conditions will vary according to the purpose of the experiment, but denaturing gels are essential if accurate molecular weights are to be determined and are commonly used prior to "Northern" transfer to nitrocellulose for hybridization experiments. There are a number of denaturing agents available, including: glyoxal, formamide, and methyl mercury (2,3). It must be remembered that all these compounds are highly toxic and should be treated with care. Their use may not always be suitable for undergraduate laboratory classes, so the basic procedure described here is concerned with nondenaturing gels. The protocol for denaturing gels using glyoxal is also given as an optional alternative, however, and may be used if desired.

Either procedure can be accomplished in under 5 h. The RNA samples are prepared for electrophoresis while the gel is setting, followed by an electrophoresis run that takes 3 h. The separated bands can be viewed on a transilluminator within half an hour of the completion of electrophoresis. The protocol described here can be conveniently used as an extension of the experiment concerned with the extraction of RNA described in Chapter 12.

2.   Materials Provided

2.1.   Solutions and Chemicals

Alcohol in wash bottles
Gel running buffer, diluted to working concentration
Agarose solution (1.1% w/v) at 50°C in 100 mL portions
Gel loading buffer
Nucleic acid samples stored as precipitates under alcohol prepared according to the procedure described in Chapter 12
Ethidium bromide solution

2.2.   Equipment

Microcentrifuge
Glass casting plate
Zinc oxide tape

Scissors
Leveling table and spirit level (optional)
Well-forming comb
50°C water bath
Electrophoresis tank and power supply
Shallow tray or plastic box, for staining gel
Disposable gloves
Safety glasses
Transilluminator
Camera, filters, camera stand, film (optional)

3. Protocol

3.1. Casting of Agarose Gel

1. Using the alcohol provided, wash the glass plate
   carefully to remove any grease. Dry the plate
   thoroughly, then form a wall around it using zinc
   oxide tape, ensuring that the tape does not pro-
   trude beyond the bottom edge of the plate. Press
   the tape firmly against the edge of the glass plate,
   particularly round the corners and where the tape
   overlaps, to ensure firm attachment. Do not
   stretch the tape or it will bow along the sides of
   the plate. The result is a leakproof wall, approximately 9 mm high, all around the plate, and is
   illustrated in Fig. 1, Chapter 3.

2. Place the prepared plate on a level surface, using
   a leveling table if provided, and place the well-
   forming comb in position about 3 cm from one
   end of the plate. Ensure that no part of the comb
   touches the glass plate.

3. Remove the agarose solution from the water bath
   and immediately pour it onto the casting plate.
   (Do not worry if there are any small leaks. They
   will rapidly seal as the agarose sets.) Ensure that
   no air bubbles are trapped under the comb, and
   leave the gel to set. This takes approximately half
   an hour, with the gel taking on a cloudy appearance when set. The nucleic acid samples can be

prepared as described in section 3.2, step 1, while the gel is setting.

3.2. Loading and Running Gel

1. Centrifuge the alcohol precipitate of the nucleic acids for 2–3 min in a microcentrifuge. Drain the tubes, discarding the supernatants (in a radioactive waste receptacle if appropriate), and dry the precipitate briefly under a stream of nitrogen or by leaving the tube upside down on a paper tissue for a few minutes. Absolute dryness is not required, but the bulk of the alcohol must be removed to allow the sample to dissolve.

2. Dissolve the nucleic acid samples in 30 μL of loading buffer. Tap the tube sharply several times, or use a vortex mixer briefly, to resuspend the pellets. The samples will take 5–10 min to dissolve, during which the electrophoresis apparatus can be assembled.

3. Carefully remove the comb from the set gel and place the casting plate and gel into the electrophoresis tank with the sample wells toward the cathode (negative electrode). The zinc oxide tape may be removed or, if desired, left in place to prevent unwanted movement of the gel; it has no effect on electrophoresis. Pour sufficient running buffer into the tank to completely submerge the gel (and tape, if present) to a depth of a few millimeters. Connect the power supply and check that a voltage of 100 V gives a current of between 40 and 60 mA. The precise current obtained will vary according to the design of the apparatus, but a very low current indicates a break in the electrical circuit.

4. Switch off the power supply and carefully load the samples into the wells using an automatic pipet or syringe. Take care not to puncture the bottom of the wells—the loading buffer is very dense and should fall easily into the wells so the syringe or pipet tip can be held a few millimeters

above the well. The wells should comfortably hold the entire 30 μL.

5. When all the samples are loaded, put the lid on the electrophoresis tank and run the gel at 100 V for 3 h or until the blue dye has run at least 8 cm.

6. To stain the gel, switch off the power supply and place the gel (with the tape removed) in a suitable tray or box. Cover the gel with used running buffer and add 200 μL of ethidium bromide solution to the side, not directly on top, of the gel. Mix in the ethidium bromide by gently rocking the tray.

7. Allow the gel to stain for at least half an hour, then, wearing rubber gloves, transfer the gel to a tray containing running buffer without ethidium bromide.

8. Place the gel directly on a UV transilluminator and cover with a Perspex screen. Put on safety glasses and view the gel by turning on the UV light. Nucleic acids on the gel will appear orange because of the fluorescence of bound ethidium bromide. Take a photograph of the gel, if desired, to keep a permanent record.

4. Denaturing Gels

4.1. Introduction

To obtain accurate molecular weights and observe the mobility of RNA in the absence of secondary structures, it is necessary to use denaturing agents during electrophoresis.

Described below are the modifications required to the protocol described in section 3 for running nucleic acids denatured by treatment with glyoxal and dimethyl sulfoxide. The basic procedure for running the gels is the same as that already described, except for the choice of running buffer; in this case, a low ionic strength, phosphate buffer is used to maintain the nucleic acids in a denatured state. The buffer is slowly consumed during electrophoresis and must be constantly circulated using a peristaltic pump, or changed at intervals during the run.

4.2.   Materials Provided

All items described in section 2
Glyoxal solution
Dimethyl sulfoxide solution (DMSO)
0.03M Phosphate buffer
Peristaltic pump (optional)

4.3.   Protocol

1. Collect the nucleic acid precipitates as described in section 3.2, step 1. Dissolve the nucleic acid sample in 6 μL 0.03M phosphate buffer, add 9 μL DMSO and 3 μL glyoxal solution. Tightly cap the tube and incubate at 50°C for 60 min.

2. Pour the agarose gel, melted in phosphate buffer, as described in section 3.1.

3. Cool the nucleic acid samples to room temperature and add 12 μL of phosphate-loading buffer.

4. Load and run the gel, submerged in phosphate buffer, as described in section 3.2. Constantly recirculate the buffer using a peristaltic pump or change the buffer every 30 min.

5. Stain and view the gel as described in section 3.2. **Note**: Staining of glyoxylated RNA is improved by deglyoxylation using alkali. Soak the gel in 300 mL 50 mM NaOH, containing 0.5 μg/mL ethidium bromide, for 20 min, then transfer the gel to 300 mL of 0.1M Tris-HCl (pH, 7.5), containing 0.5 μg/mL ethidium bromide for 40 min. Finally, transfer the gel to 0.1M Tris-HCl (pH, 7.5) and visualize the RNA with UV light.

5.   Results and Discussion

After staining, several bands should be visible on the gel, as shown in Fig. 1. DNA, if present, runs as a single band of

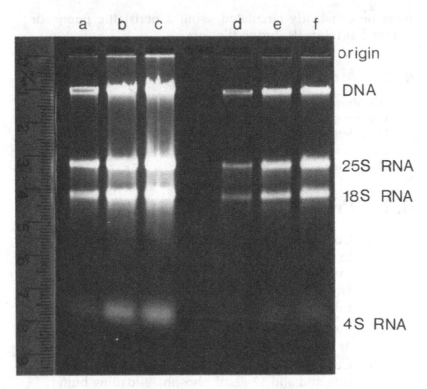

Fig. 1. A 1.1% agarose nondenaturing gel run for 3 h at 100
V. Samples a, b, and c were 2, 4, and 7 µg of total nucleic acids
from dark-grown maize shoots and samples d, e, and f were 1, 3,
and 5 µg of total nucleic acids from dark-grown pea shoots,
prepared according to the procedure described in Chapter 12. The
scale on the left shows the distance from the origin in cm. Note
that the higher loadings are necessary to observe small molecular
weight RNA.

the lowest mobility. Two or more rRNA bands should be
visible, depending on the source of nucleic acids, and small
molecular weight RNA will be observed as one, possibly
two, bands at the furthest distance from the origin.

    Measure the distance moved by each RNA band from
the front of the loading well and plot these distances
against log molecular weight using the data shown in Table
1. The plot should be linear over the 28–16S RNA range,
although nondenaturing gels are not necessarily reliable in
this respect. The standard plot can be used to determine
the molecular weights of unknown RNA.

Table 1
Molecular Weights of Ribosomal RNA

RNA	Source	Mol wt
28S	Mammal cytoplasm	1.74×10^6
25S	Plant cytoplasm	1.30×10^6
23S	Chloroplast	1.10×10^6
18S	Plant cytoplasm	0.69×10^6
16S	Chloroplast	0.56×10^6
5S	Plant cytoplasm	3.8×10^4

References

1. Thomas, P.S. (1980) Hybridization of denatured RNA and small DNA fragments transferred to nitrocellulose. *Proc. Natl. Acad. Sci. USA* **77**, 5201–5205.
2. Lehrach, H., Diamond, D., Wozney, J.M., and Boedtker, H. (1977) RNA molecular weight determination by gel electrophoresis under denaturing conditions, a critical re-examination. *Biochemistry* **16**, 4743–4751.
3. Carmichael, G.G. and McMaster, G.K. (1980) The analysis of nucleic acids in gels using glyoxal and acridine orange. *Meth. in Enzymol.* **65**, 380–391.

Table 2

Molecular Weights of Ribosomal RNA

RNA	Source	$M(\times 10^6)$

References

Chapter 14

In Vitro Protein Synthesis

Jayne A. Matthews
and Raymond A. McKee

1. Introduction

The technique of in vitro protein synthesis involves the translation of messenger RNA (mRNA) in a cell-free extract to yield polypeptide products. Both prokaryotic and eukaryotic systems have been developed, but for this experiment only a eukaryotic system will be considered. In 1973, Roberts and Paterson (3) demonstrated that a cell-free extract of wheat germ would faithfully translate rabbit globin mRNA and tobacco mosaic virus RNA to yield full-length polypeptide products. Other systems have also been developed, one of the most popular being the rabbit reticulocyte lysate developed by Pelham and Jackson (2) in 1976. Both wheat germ and reticulocyte lysates are used extensively, and both have advantages and disadvantages (Table 1), but the choice of system often depends on personal preference. For the experiment outlined in this chapter, the wheat germ lysate is used because it is easy to prepare and the starting material is readily available.

To make the lysate, wheat germ is ground in a buffer and the homogenate is centrifuged at 30,000g. The supernatant (often referred to as S30) contains ribosomes, initiation and elongation factors, tRNA, aminoacyl tRNA synthetases, and pools of nucleotides, amino acids, and other

Table 1
Advantages and Disadvantages of Wheat Germ
and Reticulocyte Lysates

Wheat Germ	Reticulocytes
Advantages	
Very low cost	˙Very high translational activity
Small pools of endogenous amino acids and ions	No premature termination occurs
Very low endogenous mRNA levels	Reaction mixture is simpler to assemble
Disadvantages	
Lower activity than reticulocyte	Expensive to buy and difficult to make
Some premature termination of polypeptides occurs	Needs treating with nuclease to destroy endogenous mRNA

small molecules. In order to standardize the translation assay, it is useful to remove the pools of small molecules, which is done using a Sephadex G25 desalting column. The ribosomes and macromolecules pass through the column in the void volume, but the small molecules are retarded. Although the process of making a wheat germ lysate is simple, a number of factors are important in obtaining an active extract. The wheat germ must not have been heat-treated (e.g., roasted), and should not contain too much white endosperm material. Sterile buffers and equipment should be used and skin contact with reagents avoided; these precautions eliminate contamination with ribonuclease. Speed and cold temperatures are essential, and storage at $-80°C$ is necessary for long-term (greater than 2 wk) maintainence of the extract.

For translation of mRNA, the lysate is supplemented with GTP, magnesium and potassium ions, amino acids, a radiolabeled amino acid, dithiothreitol (DTT), ATP (plus a regeneration system of creatine phosphate and creatine phosphokinase), and the RNA. The polyamine spermidine enhances translation, but sodium and chloride ions are inhibitory. The level of translation activity is critically dependent on the concentration of magnesium ions in the assay, the optimum generally being around 2.5 mM. This optimum must be determined for any new batch of wheat germ lysate.

Translation activity is determined by measuring the incorporation of radiolabeled amino acids into protein; this quantitation is explained more fully in the results section. In addition, the analysis of polypeptide products is of great importance. In this experiment, the products are analyzed by SDS-polyacrylamide gel electrophoresis (SDS-PAGE), and the radiolabeled products are visualized by autoradiography. This technique demonstrates the total translation products from an mRNA or viral RNA fraction, but it is often necessary to determine whether or not the synthesis of one particular protein has occurred. In this case, SDS-PAGE is only the first step, and specific proteins are detected using immunological techniques or methods that exploit unique properties of the protein (e.g., unusual solubility). Comigration of the in vitro-synthesized product with the authentic protein is not sufficient evidence, particularly since comigration often does not occur. The anomaly in electrophoretic mobility results from the fact that wheat germ lysate does not carry out secondary modifications such as glycosylation, adenylation, or proteolytic cleavage of precursors. The same limitations apply to the use of rabbit reticulocyte lysate. Methods for the detection of specific proteins are given in Chapter 17.

In vitro protein synthesis makes a number of important contributions to the technology of gene cloning. Complementary DNA (cDNA) is synthesized using mRNA as a template, and it is important to know that the mRNA population encodes the protein of interest. Translation of the mRNA in vitro is the principal method of characterizing the RNA fraction. In addition to this application, in vitro protein synthesis is also of value in identification of cDNA clones. The mRNA molecule encoding the protein of interest will hybridize to any recombinant plasmid containing the complementary sequence (i.e., the plasmid of interest). When this hybridization occurs, the mRNA molecule can no longer be translated; hence, if a mixture of mRNA and a recombinant plasmid are translated in vitro, the mRNA complementary to the plasmid will not yield a polypeptide product. The clone carrying the plasmid is therefore identified on the basis of a polypeptide missing from the translation products. This technique is known as hybrid-arrested translation (HART) (1). In the reverse of this technique, hybrid-release translation (HRT), the plasmid containing the cloned DNA is used to positively select the mRNA complementary to it; the mRNA purified in this way gives rise to just one polypeptide product. The use of these two techniques together gives an unequivocal identification of the clone.

The aim of these experiments is to demonstrate the basic technique of in vitro protein synthesis; the variations (HART and HRT) will not be attempted. A wheat germ lysate will be prepared and optimized for magnesium concentration by translation of commercially available rabbit globin mRNA. Product analysis (by SDS-PAGE and autoradiography) will be carried out on polypeptides synthesized using rabbit globin mRNA and TMV RNA.

Many other RNA preparations would be suitable, and this protocol could be combined with another in which RNA is prepared (see, for example, Chapter 12 or ref. 4). If time or equipment do not permit the carrying out of Expt. 3 (product analysis), Expts. 1 and 2 alone can be attempted. Experiments 1 and 2 can be carried out in one full-day or two half-day sessions. Experiment 3 requires a full day, and some time for subsequently processing an electrophoretic gel and autoradiograph.

2. Materials Provided

2.1. Expt. 1, Preparing the Wheat Germ Lysate

2.1.1. Materials and Solutions

Wheat germ
Mortar and pestle, 4°C
Sterile Pasteur pipet
15-mL Sterile glass centrifuge tubes capable of withstanding 30,000g, plus adaptors for 8 × 50-mL rotor
Sterile spatula

8 × 50-mL rotor, 4°C
Glass chromatography column
10-mL Syringe barrel plus a 21-gage needle or
    disposable pipet tips
Plastic universal bottles
Cooled centrifuge, 4°C
Freezer, −20°C or −80°C, or liquid N_2
Ice buckets and ice
Safety glasses
Wheat germ grinding buffer (WGGB)
Sephadex G25, and equilibration buffer
Liquid N_2

2.2. Expt. 2, Optimizing the Magnesium Concentration for the Wheat Germ Lysate

2.2.1. Materials

Micropipet, 20 μL and 100 or 200 μL
Sterile tips to fit pipets
Sterile Eppendorf tubes
150-mL beakers
Ice bucket, lid, and ice
Filter paper rectangles (1 × 2 cm, approximately)
Millipore forceps
Pins
Paper tissues
Paper towels
Glass rod
Disposable latex gloves
Waterbath, 28°C
Waterbath, boiling
Racks for waterbath, to hold Eppendorf tubes
Microcentrifuge, to spin Eppendorf tubes
Scintillation vials and caps
Aluminum foil
Hot plates
Solvent waste bottle
Scintillant dispenser
Scintillation counter
Washing up bowl and water + decon 90 detergent
Fumehood and Benchcote

2.2.2.   Solutions and Chemicals

Bottle of sterile H_2O
100 mM magnesium acetate
20% TCA and 0.5% methionine
Industrial methylated spirit (IMS):diethyl ether, 1:1
IMS
5% TCA
Energy mix:
   ATP/creatine phosphate
   Spermidine
   Amino acid solution
   DTT
   Creatine phosphokinase
   GTP
   [^{35}S]Methionine
   Hepes—KOH, potassium acetate
Globin mRNA
Wheat germ lysate

2.3.   Expt. 3, Translation for Analysis of Products

2.3.1.   Materials and Solutions

Use apparatus as for Expt. 2, in addition to the following:

SDS polyacrylamide gel (*see* Chapter 15)
Plastic boxes to stain and destain gel
Gel dryer
Sheets of Whatman 3MM paper (or equivalent)
Clingfilm
Glass plates (same size or larger than gel)
Masking tape
Aluminum foil
Black polythene bag
Freezer $-80°C$, or less effective, $-20°C$
X-ray film
Power pack
25-μL syringe
Gel stain
Gel destain
5% TCA
Cracking buffer

3. Protocol

3.1. Expt. 1, Day 1. Preparing the Wheat Germ Lysate

1. Prepare a Sephadex-G25 column in the cold room. Set the column in a retort stand and pipet into it the autoclaved Sephadex. Allow the buffer to flow through as the Sephadex settles, and fill as necessary to give a column 20 cm high. Wash the column with at least 50 mL wheat germ column buffer. Clamp off the column until you are ready to use it.

2. Weigh 8 g wheat germ, and transfer to a mortar precooled to 4°C. Break up a sterile Pasteur pipet into the mortar, and grind the wheat germ with the glass for approximately 1 min. *Wear safety glasses.*

3. Add 16 mL ice-cold wheat germ grinding buffer and continue to grind to a thick paste.

4. Transfer the paste to 2 × 15-mL centrifuge tubes using a sterile spatula. Spin at 30,000g for 20 min at 4°C in an 8 × 50-mL rotor. The speed required is 16,000 rpm on most makes of centrifuge.

5. Carefully collect the aqueous layer with a Pasteur pipet and transfer to a fresh centrifuge tube on ice. Avoid disturbing the soft surface of the pellet or the floating lipid layer.

6. Measure the volume, using a sterile pipet. Having allowed the G25 column to run almost dry, apply the wheat germ extract. Wash it through with column buffer. When the eluate appears cloudy, begin collecting it into a fresh centrifuge tube and continue collecting until you have a volume equal to the original volume of extract.

7. Spin the extract at 10,000 rpm for 10 min at 4°C to clarify. Collect the supernatant and record the volume.

8. The lysate must be frozen quickly in small aliquots. There are two simple methods, either of which can be used.

   (a) Pipet 50-μL aliquots into plastic pipet tips. Dip into liquid nitrogen to freeze, then eject the tips into the liquid nitrogen. Collect these and store at −80°C, if possible. Storage at −20°C is possible for up to 1 wk.

   (b) Set up a 10-mL syringe barrel with a 21-gage syringe needle on a retort stand over a container of liquid nitrogen. Pour the lysate into the syringe and allow it to drip through the needle. Collect the frozen spheres and store in a plastic universal bottle at −80°C or −20°C, as described above.

3.2. Expt. 2, Day 1. Optimizing the Magnesium Concentration for the Wheat Germ Lysate

Carry out all of the following procedures on ice.

1. Prepare the energy mix:

   50 μL Hepes–KOH, pH 7.6; potassium acetate

   10 μL ATP/creatine phosphate

   10 μl Spermidine

   10 μl Amino acid solution

   4 μL DTT

   4 μL Creatine phosphokinase

   4 μL GTP

   8 μL [^{35}S]Methionine

   Pipet into a 1.5-mL tube. Mix contents and spin for 10 s in a microfuge to ensure the contents are at the bottom of the tube. Store the tube on ice.

2. Prepare the following concentrations of magnesium solutions by diluting the 100 mM magnesium acetate stock with sterile water. Make 100 μL of each concentration:

   5 mM Magnesium acetate

   10 mM Magnesium acetate

   15 mM Magnesium acetate

20 mM Magnesium acetate
25 mM Magnesium acetate
30 mM Magnesium acetate

Store the tubes on ice:

3. Label eight 1.5-mL tubes with numbers 1–8, and prepare reaction mixtures according to Table 2. Mix the contents of the tubes and spin for 10 s in a microfuge.

4. Incubate the reaction mixtures at 28°C for 60 min. Return the tubes to ice at the end of the incubation period.

5. While the tubes are incubating, prepare filters for assaying the incorporation of radioactivity into protein. Label (using pencil) three sets of eight 2 × 1-cm filter paper strips with 1A, 1B, 1C; 2A, 2B, 2C; 3A, 3B, 3C; and so on. Impale these on pins and push the pins into a piece of expanded polystyrene covered in foil (an ice bucket lid is ideal), so that the filter paper strips are supported above the surface. Pipet 20 μL of 20% TCA and 0.5% methionine into the center of each filter and allow to dry.

6. From each of your reaction mixtures, pipet 5 μL onto each of the appropriately number filters, i.e., from No. 1, pipet 5 μL onto 1A, 1B, and 1C. Allow to dry for 5 min.

7. Do not wash the "A" series filters.

8. Remove the pins. Wash the "B" series in 50 mL of 5% TCA at room temperature for 10 min.

9. Wash the "C" series filters in 50 mL of 5% TCA at about 85–95°C for 10 min, then rinse in 5% TCA at room temperature.

10. Finally, rinse both the "B" and "C" series filters in 50 mL IMS:diethyl ether (1:1), and then in 50 mL diethyl ether. Spread the filters out on a sheet of aluminum foil to dry.

11. Place each filter into a scintillation vial containing 6 mL scintillant. Screw the caps on tightly, and count the samples in a liquid scintillation counter. Use the ^{14}C channel and count each sample for 2 min.

Table 2
Preparation of Reaction Mixtures for Expt. 2

Tube	Energy mix, μL	Magnesium acetate solution	Sterile H$_2$O μL	0.005 μg/mL globin mRNA, μL	Wheat germ lysate, μL	Final Mg^{2+} conc., mM
1	10	5 μL of 15 mM	15	0	10	2.5
2	10	5 μL of 0 mM	10	5	10	1.0
3	10	5 μL of 5 mM	10	5	10	1.5
4	10	5 μL of 10 mM	10	5	10	2.0
5	10	5 μL of 15 mM	10	5	10	2.5
6	10	5 μL of 20 mM	10	5	10	3.0
7	10	5 μL of 25 mM	10	5	10	3.5
8	10	5 μL of 30 mM	10	5	10	4.0

12. The "A" series filters give a measure of the total radioactivity added to the tube. The "B" series filters give a measure of the incorporation of [^{35}S]methionine into protein and aminoacyl tRNA. The heating treatment given to the "C" filters hydrolyzes amino acyl tRNA, and hence the "C" filters give a measure of the incorporation of [^{35}S]methionine into protein. Plot a graph of cpm (counts per minute) incorporated into protein against magnesium concentration. Calculate the percentage of total cpm incorporated into protein (i.e., cpm on filter C/cpm on filter A $\times$ 100), and plot this against magnesium concentration. From these results, you should be able to determine the magnesium concentration that gives optimum protein synthesis.

3.3. Expt. 3, Day 2. Translation for Analysis of Products

1. Prepare the energy mix.

   5 μL ATP/creatine phosphate
   5 μL Spermidine
   5 μL Amino acid solution
   2 μL DTT
   2 μL Creatine phosphokinase
   2 μL GTP
   9 μL [^{35}S]Methionine
   25 μL Magnesium solution
   25 μL Hepes–KOH pH 7.6, potassium acetate

   Pipet into a 1.5-mL tube. You should use the magnesium solution that gave you the maximum incorporation of methionine into protein in Expt. 2. Mix the contents and spin for 10 s. Store the tube on ice.

2. Label three tubes with numbers 1–3 and prepare reaction mixtures according to Table 3.

3. Incubate the reaction mixtures at 28°C for 60 min and return the tubes to ice after 60 min.

4. Assay the incorporation of [^{35}S]methionine into protein in exactly the same way as described for

Table 3
Preparation of Reaction Mixtures for Expt. 3

Tube	Energy mix, μL	0.005 μg/mL globin mRNA, μL	0.1 μg/mL TMV RNA, μL	Wheat germ lysate, μL	Sterile H$_2$O, μL
1	16	—	—	10	14
2	16	5	—	10	9
3	16	—	5	10	9

Expt. 2, steps 6–12. If the remainder of the experiment is not carried out on the same day, store the reaction tubes at −20°C until required.

5. There should now be 25 μL remaining in each reaction tube. Add 25 μL SDS-denaturation mix (cracking buffer) to each. Pierce a hole in the lid of each tube, and suspend them in a boiling water bath for 2 min. (The hole allows pressure to escape.) Cool the samples to room temperature.

6. The samples can now be loaded onto an SDS-polyacrylamide gel for separation of the polypeptides. Full details of this technique are given in Chapter 15. These directions should be followed exactly.

7. When the gel has been stained and destained, it must be dried in order to detect the radiolabeled polypeptides by autoradiography. Transfer the gel to a piece of Whatman 3MM filter paper wetted with distilled water and place the gel on the dryer. Follow the manufacturers instructions for drying the gel. The procedure should take about 2 h.

8. Place the dried gel on a glass plate and hold it in place with masking tape. In a dark room (illuminated only with a safe light), place a sheet of X-ray film on top of the gel and a second glass plate on top of the film. Tape the "sandwich" together with masking tape. Wrap in five layers of aluminum foil and one layer of black polythene. Place the package in a −80°C freezer (or on dry ice if a freezer is unavailable, though this is less efficient).

9. After 4–5 d, develop the X-ray film for 5 min and fix for 2–4 min. Wash the film in running water for 20 min and hang up to dry.

4. Results and Discussions

There are no results from Expt. 1 apart from recording the volume of extract prepared, but Expt. 2 yields vital data concerning the activity and characteristics of the lysate for cell-free protein synthesis. The counts per minute (cpm) obtained from the A, B, and C series filters should be tabulated, and the percentage of the total radiolabeled amino acid incorporated into protein should be calculated according to the formula:

$$\text{Percent incorporation} = \frac{\text{CPM on filter C} \times 100}{\text{CPM on filter A}}$$

The C series of filters represent incorporation of radioisotope into protein and the A series of filters give a measure of the total amount of radioisotope added to the reaction mixture. Sample results are given in Table 4. The results may also be represented graphically plotting the percent incorporation against magnesium ion concentration.

For the results in Table 4 the optimum synthesis of protein occurred at a magnesium ion concentration of 2.5 mM. The optimum usually falls between 2.0 and 2.8 mM. The actual magnitude of the counts varies considerably between the wheat germ preparations, but is usually fairly consistent between duplicate translations with the same preparation. On some occasions, a smooth optimum curve is not obtained, some points being wildly aberrant. This problem is usually caused by pipeting errors, since small volumes are often difficult to transfer accurately. Omission of the short spin before incubating the samples can also cause the same problem.

The results shown here are typical for an active wheat germ lysate preparation, with up to 80% of the radiolabeled amino acid being incorporated into hot-TCA-precipitable material, i.e., protein. The cold-TCA-precipitable material (estimated using the "B-series" filters) should always give higher cpm values than the C filters, since this is a measure of both protein- and aminoacyl-tRNA-bound amino acids. However, it is quite common for a reaction mixture (e.g., mixture 4 in Table 4) to give a lower value. There is no obvious explanation for this phenomenon, but the "B"-filter series does demonstrate that there is only a very small pool of radiolabeled aminoacyl–tRNA present in the reaction.

The translations carried out in Expt. 3 have a higher concentration of radiolabeled amino acid and therefore give higher incorporation. Sample counts are given in Table 5;

Table 4
Sample Results for Expt. 2

Reaction mixture	Magnesium acetate, mM	A filter, total cpm	B filter, cold-TCA-precipitable, cpm	C filter, hot-TCA-precipitable, cpm	Percent incorporation of radiolabeled amino acid
1 (no RNA)	2.5	96491	5042	2006	2.08
2	1.0	102315	7133	6059	5.92
3	1.5	101924	14862	12758	12.51
4	2.0	99648	62355	63743	63.96
5	2.5	98137	72194	73135	74.52
6	3.0	103477	31720	29664	28.66
7	3.5	95109	10106	8322	8.75
8	4.0	100946	6740	5629	5.58

Table 5
Sample Results for Expt. 3

Reaction mixture	RNA	A filter	B filter	C filter	Percent incorporation
1	None	251983	13841	12921	5.13
2	Globin 25 ng	245718	101905	102782	41.83
3	TMV, 250 ng	256324	216214	212767	83.01

note that the TMV RNA gives a higher total incorporation, but lower incorporation when the results are expressed on the basis of "per μg RNA." However, the more important result from Expt. 3 is the visual picture of the products produced by electrophoresis and autoradiography. A diagrammatic representation of the expected results is shown in Fig. 1. The wheat germ lysate minus exogenous RNA gives very little incorporation, and this is reflected by the lack of polypeptides visualized; just a small amount of low molecular weight material is visible. Globin mRNA yields a strong band at the molecular weight expected for globin. TMV RNA yields a variety of polypeptides with the principal product being about 110,000 mol wt. The production of this large TMV product in vitro is indicative of a good wheat germ lysate; if this product is not present and more low molecular weight components are visible, premature termination of polypeptides is likely to be occurring. This phenomenon is more common in wheat germ lysate than in reticulocyte lysate, but should not occur if the lysate has been prepared carefully from good-quality wheat germ. If a clear magnesium optimum is obtained and the large TMV product is synthesized, then it is safe to conclude that a satisfactory lysate has been prepared.

References

1. Paterson, B.M., Roberts, B.E., and Kuff, E.L. (1977) Structural gene identification and mapping by DNA, mRNA hybrid-arrested cell-free translation. *Proc. Natl. Acad. Sci. USA* **74**, 4370–4374.

Molecular weight

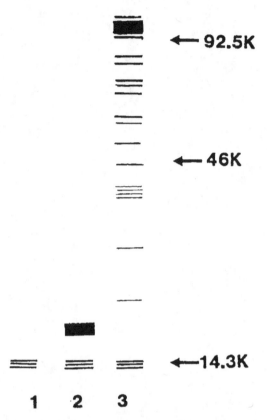

Fig. 1. Diagrammatic representation of an autoradiograph of in vitro-synthesized proteins separated by SDS-polyacrylamide gel electrophoresis. Track 1 shows endogenous protein synthesis; track 2 shows the products of globin mRNA translation, and track 3 shows the products of TMV RNA translation.

2. Pelham, H.R.B. and Jackson, R.J. (1976) An efficient mRNA-dependent translation system from reticulocyte lysate. Eur.J. Biochem. 67, 247–256.

3. Roberts, B.E. and Paterson, B.M. (1973) Efficient translation of tobacco mosaic virus RNA and rabbit globin 9S RNA in a cell-free system from commercial wheat germ. Proc. Natl. Acad. Sci. USA 70, 2330–2334.

4. Spindler, S., Siebert, P., Coffman, F., and Jurnak, F. (1984) Isolation of biological active mRNA. Biochem. Education 12, 22–25.

Chapter 15

SDS Polyacrylamide Gel Electrophoresis of Proteins

Determination of Protein Molecular Weights and High-Resolution Silver Staining

John M. Walker

1. Introduction

Polyacrylamide gel electrophoresis (PAGE) is a quick and sensitive method for analyzing the composition of mixtures of proteins. Since the early 1970s, this method has become a routine and frequently used analytical procedure in all protein chemistry laboratories, and as such, biology students should of course be familiar with this basic technique. Originally gels were formed in small glass tubes and one sample loaded on each gel, but between-run variation made it difficult to directly compare one sample with another. This problem was overcome by the introduction of the slab gel system (1), in which a large number of samples can be compared on a single run. It is this system that is described in this experiment.

There are many variations of the PAGE method, but the one that has found most common use is SDS polyacrylamide gel electrophoresis (see below for theory). SDS-PAGE separates proteins according to their size and, as a consequence, can be used to determine the molecular weight of an unknown protein mixture. Indeed, the determination of protein molecular weights by SDS gel electrophoresis is currently the preferred method of choice. Therefore, rather than just demonstrating gel electrophoretic patterns of proteins, this experiment is designed to allow the student to determine the molecular weight of an unknown protein. This examines the student's ability to generate and interpret data and of course provides results to which a mark can be assigned. The total time necessary to prepare and run a gel is about 5 h. Much of this time, however, is "passive," for example, while waiting for gels to set, and particularly while gel electrophoresis is being performed ($\sim$3 h). Depending on the total time available, it is therefore possible to run additional experiments in parallel with this one. (The determination of protein molecular weights by thin-layer gel filtration, Chapter 18, is an ideal example, since it requires little setting up and no attention while the run is being carried out. The Western blotting experiment described in Chapter 17 is also suitable for running in parallel with this experiment.)

The basic theory behind SDS gel electrophoresis is as follows: Polyacrylamide gel electrophoresis utilizes a crosslinked acrylamide support through which the protein samples are electrophoresed. The acrylamide gel is formed from acrylamide monomer and 'bis'-acrylamide, which provides the crosslinking between monomer chains (Fig. 1). The polymerization process is a free-radical chain reaction, initiated by ammonium persulfate and N,N,N',N'-tetramethylenediamine (TEMED), which are present in the polymerization mixture. The persulfate activates the TEMED, leaving it with a reactive unpaired electron. This radical then reacts with an acrylamide monomer to produce a new radical, which in turn reacts with another acrylamide molecule (or occasionally a bis acrylamide molecule), the process continuing to build up a crosslinked polymer. The

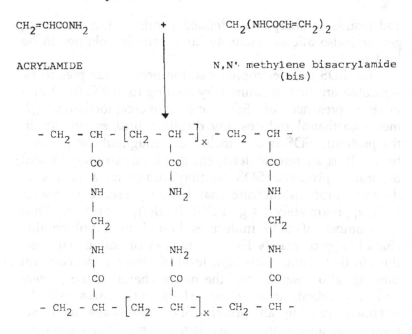

Fig. 1. The polymerization of acrylamide.

amount of crosslinking in the gel controls the "pore size" in the gel and consequently determines the molecular weight range of the proteins that can be separated in a given gel. For example, in a highly crosslinked gel, small molecular weight proteins can enter the gel matrix, but large molecular weight proteins cannot. The polyacrylamide content of a gel can vary from 3 to 30%, where the percentage value is the total acrylamide content (monomer *and* bis acrylamide, w/v.) The gel system used in this experiment is a 15% gel. The gel slab is formed in two parts. The main part of the slab comprises a 15% *separating* gel and, as the name suggests, it is in this gel that proteins are separated. On top of this separating gel is a short, low-percentage *stacking* gel into which loading wells are formed with a comb when this gel is polymerizing. Each protein sample is loaded into a well and the electrophoresis of these samples through this stacking gel results in the sharpening or "stacking" of each protein sample before it enters the separating gel. This stacking system employs the principle of isotachophoresis and results in sharp, well-defined bands in the separating gel, and also allows relatively large sample volumes to be loaded.

For SDS gel electrophoresis, the protein samples to be separated are first denatured by heating to 100°C for 2 min in the presence of SDS and β-mercaptoethanol (β-mercaptoethanol reduces any disulfide bridges present in the protein). SDS is also included throughout the gel system. SDS is an anionic detergent that binds strongly to, and denatures, proteins. SDS binding studies on a variety of different proteins indicate that in the presence of excess SDS, approximately 1.4 g of SDS binds/g of protein. Thus the number of SDS molecules bound to a polypeptide chain is approximately half the number of amino acid residues in that chain. This high level of binding and constant binding ratio 'swamps-out' the native charge of the protein and a constant negative net charge/unit mass will be obtained, causing all protein–SDS complexes to travel toward the anode on polyacrylamide gels. When protein–SDS complexes are run on polyacrylamide gels, different sized proteins show different mobilities because of the sieving effect of the gel. The separation of proteins treated with SDS is, therefore, a reflection of the difference in sizes of the proteins. It is for this reason that SDS gel electrophoresis offers a convenient method for determining the molecular weight of an unknown protein. The molecular weight of a protein under investigation may be calculated by comparing its electrophoretic mobility with those of proteins of known molecular weight (2). Using known standards, a plot of log molecular weight against mobility can be derived. Knowing the mobility of the unknown protein, its molecular weight can be read from the graph.

The Coomassie stain is used in this experiment to stain gels, and can identify as little as 0.1 μg of protein in a gel band. Although this is sufficiently sensitive for most purposes, there are occasions when even greater sensitivity is required and this greater sensitivity can be achieved by using a silver stain. Silver staining procedures have been developed from a variety of histochemical and photochemical techniques, although the precise chemistry of the method is poorly understood. Such a stain is at least 100 times more sensitive than the Coomassie stain and is comparable in sensitivity to autoradiography of radiolabeled proteins. A method for silver staining gels is described in this experiment as an additional, optional procedure. The method is based on the procedure originally described by Morrissey (3). Since this is simply a demonstration of a staining procedure, it is convenient for four students to stain a single gel between them. The method is best carried out on a gel previously stained with Coomassie Brilliant Blue.

2. Materials Provided

2.1. SDS Gel Electrophoresis

Stock acrylamide solution
1.875M Tris-HCl buffer, pH 8.8
0.6M Tris-HCl buffer, pH 6.8
10% (w/v) SDS
5% (w/v) Ammonium persulfate
N,N,N',N'-tetramethylenediamine (TEMED),
    sited in a fume cupboard
Electrode buffer
Protein samples in SDS sample buffer
Protein stain solution
Destain solution
Box or dish for staining/destaining
Power pack and electrophoresis apparatus
Microsyringe or automatic pipet
100-mL Buchner flask and bung
Methylated spirit and tissues
Light box

2.2. Silver Staining

10% (v/v) Glutaraldehyde (SLR)
5 µg/mL Dithiothreitol
0.1% (w/v) Silver nitrate
Developer solution
2.3M Citric acid

3. Protocol

3.1. SDS Gel Electrophoresis

Acrylamide is reported to be a neurotoxin, and care should be taken when handling acrylamide solutions. Polymerized acrylamide is not toxic.

1. Join gel plates together to form the gel cassette, and clamp it in a vertical position. The exact manner of forming the cassette will depend on the type of design being used.

2. Mix the following in a 100-mL Buchner flask:

8.0 mL	1.875M Tris-HCl, pH 8.8
11.4 mL	Water
20.0 mL	Stock acrylamide
0.4 mL	10% SDS
0.2 mL	5% Ammonium persulfate

   "Degas" this solution under vacuum for about 30 s. Some frothing will be observed and one should not worry if some froth is lost down the vacuum tube. This degasing step is necessary to remove dissolved air from the solution, since oxygen can inhibit the polymerization step. Also, if the solution has not been degased to some extent, bubbles will form in the gel during polymerization, which will ruin the gel.

3. Add 10 µL of TEMED and gently swirl the flask to ensure even mixing. The addition of TEMED will initiate the polymerization reaction, and although it will take about 20 min for the gel to set, this time can vary depending on room temperature, so it is advisable to work fairly quickly at this stage.

4. Using a Pasteur (or larger) pipet, transfer this separating gel mixture to the gel cassette by running the solution carefully down one edge between the glass plates. Continue adding this solution until it reaches a position 3 cm from the top of the cutaway gel plate.

5. To ensure that the gel sets with a smooth surface, carefully run distilled water down one edge into the cassette using a Pasteur pipet. Because of the great difference in density between the water and the gel solution, the water will spread across the surface of the gel without serious mixing. Continue adding water until a layer of about 2 mm exists on top of the gel solution.

6. The gel can now be left to set. As the gel sets, heat is evolved and can be detected by carefully touching the gel plates. When set, a very clear refractive index change can be seen between the polymerized gel and overlaying water.

7. While the separating gel is setting, prepare the following stacking gel solution:

   1.0 mL   0.6M Tris-HCl, pH 6.8
   1.35 mL   Stock acrylamide
   7.5 mL   Water
   0.1 mL   10% SDS
   0.05 mL   5% Ammonium persulfate

   Degas this solution as before.

8. When the separating gel has set, pour off the overlaying water. Add 10 μL of TEMED to the stacking gel solution and use some ($\sim$2 mL) of this solution to wash the surface of the polymerized gel. Discard this wash, then add the stacking gel solution to the gel cassette until the solution reaches the cutaway edge of the gel plate. Place the well-forming comb into this solution and leave to set. This will take about 30 min. Refractive index changes around the comb indicate that the gel has set. It is useful at this stage to mark the positions of the bottoms of the wells on the glass plates with a marker pen.

9. Carefully remove the comb from the stacking gel, remove any spacer from the bottom of the gel cassette, and assemble the cassette in the electrophoresis tank. Fill the top reservoir with electrophoresis buffer and look for any leaks from the top tank. If there are no leaks, half fill the bot-

tom tank with electrophoresis buffer, tilt the apparatus to dispel any bubbles caught under the gel, and carefully continue filling the lower tank.

10. Samples can now be loaded onto the gel. Fifteen μL samples should be loaded. Place the syringe needle through the buffer and locate it just above the bottom of the well. Slowly deliver the sample into the well. The dense sample solvent ensures that the sample settles to the bottom of the loading well. Continue in this way to fill all the wells with unknowns or standards, and record the samples loaded.

11. The power pack is now connected to the apparatus and a current of 40–50 mA passed through the gel (constant current). In the first few minutes the samples will be seen to concentrate as a sharp band as it moves through the stacking gel. (It is actually the bromophenol blue that one is observing, not the protein, but of course the protein is stacking in the same way.) Continue electrophoresis until the bromophenol blue reaches the bottom of the gel. This will take 2.5–3 h.

12. Disconnect the power supply, pour off the buffer, remove the gel plates, and using a spatula, pry the gel plates apart. The gel itself may now be removed by hand (the rather sticky stacking gel can be removed and discarded since it is no longer required), and placed in a suitable container together with protein stain (~200 mL). The gel should then be gently shaken for at least 2h, and preferably overnight.

13. Pour off the protein stain and replace with destain solution (~200 mL). Continue shaking the gel. Some protein bands will already be visible and more will become apparant as the gel destains. The destaining solution should be changed once after a few hours.

14. View the destained gel on a light box and record any 'necessary observations. A typical gel is shown in Fig. 2.

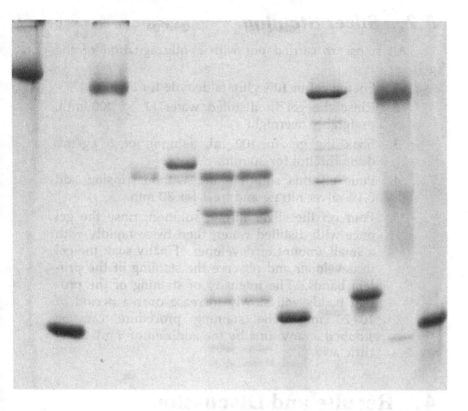

Fig. 2. A typical SDS gel, prepared by second-year biology students in a practical class. Samples loaded (15 µg each of the standard proteins) are given below.

1. Transferrin
2. Lysozyme
3. Bovine serum albumin
4. Glyceraldehyde-3-phosphate dehydrogenase
5. Lactate dehydrogenase
6. Milk
7. Milk
8. Myoglobin
9. Catalase
10. β-Lactoglobulin
11. IgG Light chain
12. Cytochrome C

3.2. Silver Staining

All steps are carried out with gentle agitation of the gel.

1. Fix the gel in 10% glutaraldehyde for 30 min.
2. Rinse the gel in distilled water (3 × 200 mL), preferably overnight.
3. Soak the gel in 100 mL solution of 5 µg/mL dithiothreitol for 30 min.
4. Pour off this solution and without rinsing add 0.1% silver nitrate and treat for 30 min.
5. Pour off the silver nitrate solution, rinse the gel once with distilled water, then twice rapidly with a small amount of developer. Finally soak the gel in developer and observe the staining of the protein bands. The intensity of staining of the protein bands will slowly increase over a period of 10–20 min. The staining procedure can be stopped at any time by the addition of 5 mL 2.3M citric acid.

4. Results and Discussion

When your gel has destained, measure the distance moved by each standard protein in the separating gel, and plot a calibration curve of log molecular weight against distance moved. It is important to note that under the denaturing conditions used, proteins are broken down into subunits. Lactate dehydrogenase, for example, consists of four equal subunits, and has a molecular weight of 140,000. However, on SDS gel electrophoresis, it has an apparent molecular weight of 35,000 since one is observing the individual subunits, and this is the figure one should use when constructing the graph. Other such examples are indicated in Table 1. Now measure the distance moved by the unknown protein (or proteins if your unknown is a mixture), and from the graph calculate the molecular weight of the unknown(s). Although the log molecular weight/mobility gives a fairly good straight line plot, there is some scatter

Table 1
Some Commercially Available Proteins of
Known Molecular Weight

Protein	Molecular weight
Lactate dehydrogenase	140,000
	(4 × 35,000)
Phosphorylase b	92,500
Transferrin	78,000
Bovine serum albumin	66,300
Catalase	50,000
Ovalbumin	43,000
β-Lactoglobulin	36,800
	(2 × 18,400)
Glyceraldehyde-3-phosphate dehydrogenase	34,000
Pepsin	34,700
Trypsinogen	24,000
Myoglobin	17,800
Lysozyme	14,400
Cytochrome C	12,400

around this line, particularly at very high and very low molecular weight values. There is therefore some degree of flexibility in drawing the graph (the method has an error of about ±10%), and your final result quoted should indicate this. Comment on the purity of the standard proteins, and compare the individual gel patterns before and after silver staining.

References

1. Laemmli, U. K. (1970) Cleavage of structural proteins during the assembly of the head of bacteriophage T4. *Nature* **227**, 680–685.
2. Weber, K. and Osborne, M. (1969) The reliability of molecular weight determinations by sodium dodecyl sulfate polyacrylamide gel electrophoresis. *J. Biol. Chem.* **244**, 4406.
3. Morrissey, J. (1981) Silver stain for proteins in polyacrylamide gels. A modified procedure with enhanced uniform sensitivity. *Anal. Biochem.* **117**, 307–310.

4. Studier, F. W. (1973) Analysis of bacteriophage T7 early RNAs and proteins on slab gels. *J. Mol. Biol.* **79**, 237–248.

Further Reading

Smith, I. (1976) *Zone Electrophoresis, Chromatographic, and Electrophoretic Techniques*, Vol. 2, Heinemann Medical Books, London.

Smith, B. J. and Nicholas, R. H. (1983) Analytical Electrophoretic Techniques in Protein Chemistry, in *Techniques in Molecular Biology* (J. M. Walker and W. Gaastra, eds.) Croom Helm, Beckenham, Kent, England.

Chapter 16

Isoelectric Focusing (IEF) of Proteins in Thin-Layer Polyacrylamide Gels

John M. Walker

1. Introduction

Isoelectric focusing (IEF) is an electrophoretic method for the separation of proteins, according to their isoelectric points (pI), in a stabilized pH gradient. The method involves casting a layer of polyacrylamide gel containing a mixture of carrier ampholytes (low-molecular-weight synthetic polyamino–polycarboxylic acids). When an electric field is applied to such a gel, the carrier ampholytes arrange themselves in order of increasing pI from the anode to the cathode. Each carrier ampholyte maintains a local pH corresponding to its pI and thus a uniform pH gradient is created across the gel. If a protein sample is applied to the surface of the gel, it will also migrate under the influence of the electric field until it reaches the region of the gradient where the pH corresponds to its isoelectric point. At this pH, the protein will have no net charge and will therefore become stationary, or "focused" at this point. Proteins are therefore separated according to their charge, and not size as with SDS gel electrophoresis. In practice the protein samples are loaded onto the gel before the pH gradient is formed. When a current is applied, protein migration and pH gradient formation occur simultaneously.

In this experiment the students will prepare their own isoelectric focusing gel capable of separating proteins with isoelectric points in the pH range of 3.5–10.0. (This pH range covers the pI range of the majority of proteins.) The gel is a low-percentage acrylamide gel (thereby eliminating any molecular sieving effect of the gel) and contains added ampholytes. The gel is polymerized by photopolymerization using riboflavin as the catalyst. Ammonium persulfate, the usual catalyst used in forming acrylamide gels, is not used because it can interfere with the isoelectric focusing. The polymerization process takes about 3 h. If this time scale is inconvenient, gels may be prepared on a previous occasion and stored at 4°C for many months. Alternatively, prepared isoelectric focusing gels can be purchased from commercial sources, although this approach is less instructive to the students. Traditionally, 1–2 mm thick isoelectric focusing gels have been used by research workers, but the relatively high cost of ampholytes makes this a fairly expensive procedure if a number of gels are to be run. However, the introduction of thin-layer isoelectric focusing (where gels of only 0.15 mm thickness are prepared, using a layer of electrical insulation tape as the "spacer" between the gel plates) has considerably reduced the cost of preparing IEF gels, and such gels are therefore used in this experiment.

The actual electrophoresis takes about 3 h, and up to 20 samples (5–10 μg of protein) can be loaded onto each gel. A variety of protein samples can be loaded, but it is informative for the students to provide some of their own samples (e.g., serum or saliva). Since considerable heat is generated during the focusing process, an electrophoresis tank with a water-cooled plate is necessary: overheating of the gel will result in distorted bands. When the gel is focused, the separated proteins are fixed by precipitation with fixing solution. This step is necessary since it washes out the ampholytes from the gel that would otherwise stain with the protein stain and mask the separated bands. Following the fixing step, the gel is treated with a protein stain, destained, and the focused proteins observed. The fixing, staining, and destaining process takes about 40 min.

2. Materials Provided

Glass plates
PVC electrical insulation tape
Stock acrylamide solution
Riboflavin solution
Ampholyte solution
Plastic spillage tray
Sample application squares
Electrode wicks
Aqueous solutions of protein samples
Fixing solution
Protein stain solution
Protein destain solution
Perspex box for staining/destaining
Power packs
Light box
Phosphoric acid solution and sodium hydroxide solution for
   wetting electrode wicks

3. Protocol

1. Thoroughly clean the surfaces of two glass plates,
    first with detergent and then methylated spirit. It
    is essential for this experiment that the glass plates
    are clean.

2. To prepare the gel mould, stick strips of insula-
    tion tape, 0.5 cm wide, around the edges of one
    glass plate. Do not overlap the tape at any stage.
    Small gaps at the corner joins are acceptable.

3. To prepare the gel solution, mix the following:

    9 mL   Acrylamide solution
    0.4 mL   Ampholyte solution
    60 μL   Riboflavin solution

    Note: Since acrylamide monomer is believed to
    be neurotoxin, steps 4 to 6 must be carried out
    wearing protective gloves.

4. Place the glass mould in a spillage tray and
    transfer the gel solution with a Pasteur pipet
    along one of the short edges of the glass mould.

The gel solution will be seen to spread slowly toward the middle of the plate.

5. Take the second glass plate and place one of its short edges on the taped edge of the mould, adjacent to the gel solution. Gradually lower the top plate and allow the solution to spread across the mould. Take care not to trap any air bubbles. If this happens, raise the top plate to remove the bubble and then lower it again. To avoid wasting solutions at this stage, it may be advisable to initially try this method using distilled water.

6. When the two plates are together, press them together firmly and discard the excess acrylamide solution that has spilled into the tray.

7. Place the gel mould on a light box and leave for 3 h to allow polymerization. Gel moulds may be stacked at least three deep on the light box during polymerization. Polymerized gels may then be stored at 4°C for at least 2 mo, or used immediately. If plates are to be used immediately, they should be placed at 4°C for ~15 min; this makes the separation of the plates easier.

8. Place the gel mould on the bench and remove the top glass plate by inserting a scalpel blade between the two plates and carefully twisting the blade. Occasionally, the gel will stick to the top plate, but the top plate plus gel can of course just as easily be used for electrofocusing. Do not bother to remove the insulation tape.

9. Carefully clean the underneath of the gel plate and place it on the cooling plate of the electrophoresis tank. Cooling water at 4–10°C should be passed through the cooling plate.

10. Down the full length of each of the longer sides of the gel lay electrode wicks, uniformly saturated with either 1.0M phosphoric acid (anode) or 1.0M sodium hydroxide (cathode). The wicks are 0.6 cm wide strips of Whatman No. 17 filter paper.

11. Samples are loaded by laying filter paper squares (0.5 × 0.5 cm), wetted with the protein sample, onto the gel surface. Leave 0.5 cm gaps between

each sample. The filter papers are prewetted with 5–8 µL of sample and applied at the center of the gel across the width of the gel.

12. Saliva samples may be applied by touching the tongue with a prewetted (distilled water) filter paper square. For blood samples, prewet the filter paper with distilled water, then let a small drop of blood touch the paper. Blood cells will lyse upon coming into contact with the water. Place the filter paper on the gel with the side touched by the blood uppermost (the filter paper will act as a filter, preventing much of the cell debris from reaching the gel surface). An alternative and more reproducible procedure is to collect a drop of blood by finger puncture, measure a small volume using a microsyringe, and dilute with 20–30 vol of distilled water. This procedure should be carried out quickly since the blood can clot in the syringe.

13. When all the samples are loaded, place a platinum electrode on each wick. Some commercial apparatus employ a small perspex plate along which the platinum electrodes are stretched and held taut. Good contact between the electrode and wick is then maintained by applying a weight to the perspex plate.

14. Apply a potential difference of 500 V across the plate. This should give a current of about 4–6 mA. After 10 min, increase the current to 1000 V and then to 1500 V after another 10 min. A current of ~4–6 mA should be flowing in the gel at this stage, but this will slowly decrease with time as the gel focuses.

15. When the gel has been running for about 1 h, turn off the power and carefully remove the sample papers with a pair of tweezers. Most of the protein samples will have come off the papers and be in the gel by now. When using thicker (1–2 mm) gels, sample application strips are routinely removed at this stage. However, when using thin-layer gels, some workers do not remove the

application strips at this stage since it is easy to damage the ultra-thin gel. This step may therefore be considered optional. Continue with a voltage of 1500 V for a further 2-2.5 h (total electrophoresis time, 3-3.5 h). During the period of electrophoresis, colored samples (myoglobin, cytochrome C, hemoglobin in the blood sample, and so on) will be seen to move through the gel and focus as sharp bands.

16. At the end of 3–3.5 h of electrophoresis, a current of about 0.5 mA will be detected. Remove the gel plate from the apparatus and gently wash it in fixing solution (200 mL) for 20 min (over-vigorous washing can cause the gel to come away from the glass plate). Some precipitated protein bands should be observed in the gel at this stage. Pour off the fixing solution and wash the gel in destain solution (100 mL) for 5 min and discard this solution. Add protein stain and gently agitate the gel for about 10 min. Pour off and discard the protein stain and wash the gel in destain solution. Stained protein bands in the gel may now be visualized on a lightbox and a permanent record may be obtained by simply leaving the gel to dry out on the glass plate overnight. A typical thin-layer isoelectric focusing gel is shown in Fig. 1.

4. Results and Discussion

The purpose of this experiment is essentially to demonstrate the technique of isoelectric focusing. However, some data can be derived from the final results. If a suitable surface pH electrode is available, the pH gradient in the gel may be plotted and the pI value for the individual proteins determined. If this method is used, it should of course be done immediately after the run, prior to fixing. Since this procedure can take some time, it is advisable to refocus the gel for a further 15 min before fixing in case any diffusion of samples has occurred. Alternatively, proteins of known pI can be run as standards and a graph of pI against dis-

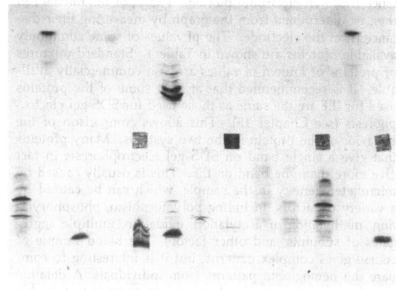

1 2 3 4 5 6 7 8 9 10 11 12 13

Fig. 1. Photograph showing part of a thin-layer isoelectric focusing gel. Samples (5 μg each) are:

1. Transferrin
2. Myoglobin
3. Bovine serum albumin
4. Sample not visible
5. Thyroglobulin (3–4 isoenzymes)
6. Human blood (1:30 dilution with distilled water)
7. Milk (1:40 dilution with distilled water)
8. α-Amylase (from *Bacillus subtilis*)
9. Catalase (from bovine liver)
10. Urease (crude Jack Bean extract)
11. Transferrin
12. Myoglobin (from sperm whale)
13. Bovine serum albumin

The cathode is at the top of the figure; the anode at the bottom. Sample application papers have been left on the gel on tracks 5, 8, 11, and 13. The pH range is from 3.5 to 10.0. The gel was stained for protein with Coomassie Brilliant Blue.

tance (from an electrode) plotted. The pI of unknowns can then be determined from the graph by measuring their distance from the electrode. The pI values of some commonly available proteins are shown in Table 1. Standard mixtures of proteins of known pI values are also commercially available. It is recommended that at least some of the proteins used for IEF are the same as those used for SDS gel electrophoresis (see Chapter 15). This allows comparison of the behavior of the protein in the two systems. Many proteins that give a single band on SDS gel electrophoresis in fact give more than one band on IEF. This is usually caused by microheterogeneity in the sample, which can be caused by a variety of factors, including polymorphism, phosphorylation, methylation or acetylation, oxidation, multiple aggregates of subunits, and other factors. The blood sample of course gives complex patterns, but it is interesting to compare the hemoglobin patterns from individuals. A detailed description of the possible hemoglobin patterns is available in ref. 1.

In recent years isoelectric focusing has been combined with SDS gel electrophoresis to give a two-dimensional technique that provides a powerful tool for the analysis and detection of proteins from complex biological sources (2). Proteins are separated according to their isoelectric points

Table 1
pI Values of Some Commonly Available Proteins[a]

Protein	pI
Methyl red[b]	3.75
Glucose oxidase	4.15
Soya bean trypsin inhibitor	4.55
β-Lactoglobulin A	5.20
Insulin (bovine)	5.35
Bovine carbonic anhydrase B	5.85
Human carbonic anhydrase B	6.55
Horse myoglobin—acidic band	6.85
Horse myoglobin—basic band	7.35
Met myoglobin (sperm whale)	8.20
Trypsinogen	9.30

[a] See also ref. 3.

[b] This is a dye and its position should be measured prior to staining.

by isoelectric focusing in the first dimension (in rod gels) and according to molecular weight by SDS gel electrophoresis in the second dimension (a slab gel). Since these two parameters are unrelated, it is possible to obtain an almost uniform distribution of protein spots across a two-dimensional gel. Protein mixtures are usually studied from tissues/organisms grown in the presence of radiolabeled amino acids, and since a protein containing as little as 1 dpm of either ^{14}C or ^{35}S can be detected by autoradiography, a protein that constitutes as little as $10^{-4}\%$ of the total protein can be detected. As many as 1800 proteins have been separated on a single gel. The reproducibility of the method is such that each spot on one separation can be matched with a spot on a different separation. Because of the high resolution and sensitivity of this technique, it has been applied to a great number of biological systems, including the analysis of proteins in tissues under various hormonal states and the analysis of proteins in tissues at different stages of development. Since the method can detect proteins that contain single amino acid substitutions that confer a change in the isoelectric point of a protein, it has been used to identify mis-sense mutations within proteins and to detect mistranslated proteins.

References

1. Detection of abnormal hemoglobins by analytical thin-layer electrofocusing in polyacrylamide gel. Application Note, 307, LKB, Sweden.
2. O'Farrell, P.H. (1975) High resolution two-dimensional electrophoresis of proteins. *J. Biol. Chem.* **250**, 509–516.
3. Righetti, P.G. and Caravaggio, T. (1976) Isoelectric points and molecular weights of proteins, a table. *J. Chromatog.* **127**, 1–28.

Further Reading

Isoelectric Focusing: Principles and Methods. 160-page book supplied by Pharmacia Fine Chemicals, AB, Uppsala, Sweden.

Righetti, P.G. and Drysdale, J.W. (1974) Isoelectric focusing in gels. *J. Chromatog.* **98**, 271-321.

Chapter 17

Protein (Western) Blotting with Immunodetection

John M. Walker

1. Introduction

Protein blotting simply involves transferring separated protein bands from an acrylamide gel onto a more stable and immobilizing medium, such as nitrocellulose paper. Once transferred to the immobilizing medium, a variety of analytical procedures may be carried out on the proteins that would otherwise have proved difficult or impossible in the gel. Such procedures may include hybridization with labeled DNA or RNA probes, detection with antibodies (as used in this experiment), detection by specific staining procedures, autoradiographic assay, and so on. The approach used here is exactly analogous to the method used to transfer DNA from agarose gels (the Southern Blot, *see* Chapter 5) and has been given the name "Western" blotting (1–3). There are a number of advantages to working with a protein blot rather than the original gel. These include:

1. Rapid staining/destaining: once transferred, staining and destaining can be achieved within 5 min.
2. There is no ampholyte staining when analyzing IEF gels.

171

3. Samples in preparative gels may be located rapidly by briefly blotting the gel.

4. Low concentrations of samples can be detected since they are not spread throughout the thickness of a gel, but "concentrated" at the surface of the paper. As a consequence, it is possible to use lower specific activities and lower concentrations of expensive or scarce probes.

5. The "blot" is a convenient permanent record of the gel run.

There are two ways of carrying out a protein blot— either by electroblotting or capillary blotting. in the method of electroblotting, proteins are transferred from the gel to the immobilizing medium by electrophoresis. This has the advantage of reducing transfer time to between 30 min and a few hours. However, since electroblotting requires relatively expensive custom-built electrophoresis apparatus, it is not used in this experiment. Instead, the alternative method of capillary blotting is used.

For capillary blotting (see Fig. 1), two or three sheets of absorbant filter paper are saturated with transfer buffer and connected to buffer reservoirs. The gel to be blotted is then placed on top of the filter paper and the transfer paper (nitrocellulose) placed on top of the gel. Two or three layers of dry blotting paper, several layers of paper towels (or disposable diapers), and a weight are then stacked above the transfer paper and gel. Blotting is then allowed

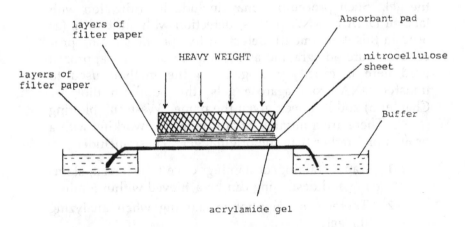

Fig. 1. A typical arrangement for capillary blotting.

to take place for 24–48 h, during which time the protein bands are carried from the gel to the nitrocellulose paper where they bind strongly. Following blotting, excess binding sites are then blocked by treating the paper with an appropriate blocking agent (e.g., Tween 80). The transferred proteins may then be investigated by using any of the methods described above.

In this experiment a freshly run SDS gel, containing replicate samples of bovine serum, is blotted for 24 h. Blotting need only be carried out on part of a gel, say four tracks, so that the rest of the gel containing different samples may be stained/destained as usual. This method may therefore be easily combined with the SDS gel electrophoresis experiment (Chapter 15). Following blotting, the nitrocellulose paper is cut in half, and one half is stained for protein. This simply demonstrates that protein has indeed been transferred to the nitrocellulose paper. Since not all the protein is transferred from the gel in the 24-h blotting period, the gel can also be stained for protein as described in Chapter 15, and the pattern of protein bands on the gel and nitrocellulose paper compared. The second half of the nitrocellulose paper is used to detect a specific protein (bovine serum albumin) within these fractionated proteins. First, the nitrocellulose paper is washed with Tween 80 to block remaining binding sites on the nitrocellulose, then the paper is incubated with a solution of rabbit anti-bovine serum albumin (first antibody). This antibody will bind to any albumin on the nitrocellulose sheet. When excess antibody has been removed by repeated saline washes, the paper is incubated with a solution of goat anti-rabbit IgG, which is conjugated to either horseradish peroxidase or alkaline phosphatase (second antibody). This second antibody will bind to any first antibody-albumin complex present on the paper. Following repeated saline washes to remove excess antibody, the paper is placed in an appropriate substrate solution for the conjugated enzyme. Substrates are chosen that have a colored end-product. The appearance of a colored band on the nitrocellulose sheet therefore determines the position of the bovine serum albumin on this sheet. Comparison of this nitrocellulose sheet with either the nitrocellulose sheet stained for protein, or the original stained gel, allows the albumin band to be

_____ Nitrocellulose sheet

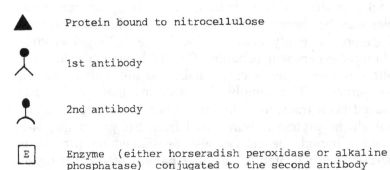

▲   Protein bound to nitrocellulose

       1st antibody

       2nd antibody

E   Enzyme (either horseradish peroxidase or alkaline
       phosphatase) conjugated to the second antibody

Fig. 2. A schematic representation of the double-antibody
technique used to identify proteins on nitrocellulose.

identified within the separated protein pattern. This
procedure is shown in diagrammatic form in Fig. 2. Once
protein blotting is completed, the nitrocellulose sheet may
be analyzed immediately or stored indefinitely, until a suitable time is available for analysis. The immunological
detection method takes approximately 2.5 h to perform.

The application of protein blotting with immunological
detection is typified by the following example. It is often
necessary to isolate the mRNA coding for a particular protein. During the isolation procedure a number of mRNA-
containing fractions are produced. Each of these must be
examined to determine which one contains the mRNA of
interest. This is normally done by adding an aliquot of
each mRNA fraction to separate in vitro protein-
synthesizing systems (see Chapter 14) and examining the
proteins produced by gel electrophoresis. Complex mRNA
mixtures will obviously synthesize a range of proteins, and
even if the protein of interest is run as a standard on the

same gel, it is not always easy to determine whether this protein is present in the incubation mixture. Just because a protein band in the mixture runs in the same place as the standard, this does not necessarily mean that the two proteins are identical; they may simply have the same molecular weight. However, because antibodies have the ability to detect *specific* proteins, the additional steps of protein blotting followed by the double-antibody detection method described above will allow the presence or absence on the gel of the specific protein in question to be determined.

2. Materials Provided

Blotting buffer
Nitrocellulose paper with a sharp scalpel blade for cutting
Whatman No. 1 filter paper
Whatman 3MM filter paper
Disposable diaper or a number of paper towels
A heavy weight, such as a lead brick or a 1-L conical flask
    filled with water
Small dishes for staining/destaining/washing
Protein staining solution (nigrosin)
Destain solution
2% Solution of Tween 80 in PBS
Solution of rabbit anti-bovine serum albumin (first antibody)
Solution of anti-rabbit IgG conjugated to either horseradish
    peroxidase or alkaline phosphatase (second antibody)
Necessary reagents for preparing the substrate solution for
    either horseradish peroxidase or alkaline phosphatase,
    depending on which conjugate is being used
Phosphate buffered saline (PBS)

3. Protocol

1. If only part of a gel is to be used (say four tracks)
    carefully cut the gel with a scalpel blade.
2. Assemble five layers of Whatman No. 1 filter
    paper, as shown in Fig. 1, and thoroughly wet
    with blotting buffer. Carefully lay the gel to be

blotted (without its stacking gel) on this wetted filter paper. The filter paper must of course be at least as wide as the gel being blotted.

3. Cut (using a scalpel, not scissors) a piece of nitrocellulose paper exactly the size of the gel to be blotted and wet it by floating on blotting buffer. Carefully layer this wetted nitrocellulose paper over the gel, taking care not to trap any air bubbles. If the surface of the gel has dried out at all, it is easier to wet the top of the gel with blotting buffer before layering the nitrocellulose paper.

4. On top of the nitrocellulose paper, lay five layers of Whatman 3MM filter paper, cut to the same size as the gel, followed by a pad of absorbant material (e.g., disposable diaper) also cut to the same size as the gel. To the top of the sandwich, add a heavy weight and leave for 24 h. Note: Care must be taken that the absorbant materials above the gel do not in any way make contact with the lower layer of wetted filter paper, otherwise the flow of buffer will bypass the gel rather than pass *through* the gel.

5. After blotting for 24 h, dismantle the sandwich and recover the nitrocellulose paper. This may then be processed directly as described below, or kept indefinitely until required for analysis, by storing it pressed between two layers of filter paper.

6. Cut the nitrocellulose sheet in half using a scalpel blade. Stain one half for protein by placing it in nigrosin and shaking for 15 min, followed by gentle agitation in repeated changes of destaining solution. The proteins that have been transferred to the nitrocellulose paper will appear as black bands on a grey/white background. The remaining gel may also be stained for protein with Coomassie Brilliant Blue, as described in Chapter 15.

7. Place the second half of the nitrocellulose filter in a small plastic dish and wash it with gentle shaking in 2% Tween solution (20 mL) for 30 min. This will block remaining binding sites on the

nitrocellulose sheet. In all washing steps, take care that the nitrocellulose sheet is mobile in the washing solution, and not stuck to the bottom of the plastic dish. All incubation steps are carried out at room temperature.

8. Pour off the Tween solution and incubate the nitrocellulose sheet in a solution of first antibody (20 mL) for 30 min.

9. Remove the first antibody solution and wash the nitrocellulose sheet (4 × 5 min) in PBS (4 × 20 mL). This should remove all traces of free first antibody.

10. Incubate the nitrocellulose sheet in second antibody (10 mL) for 30 min, then remove this solution and wash the sheet (4 × 5 min) in PBS (4 × 20 mL). Finally, wash the sheet in the buffer used for the enzyme substrate (20 mL) for 5 min.

11. Pour off the buffer solution and wash the sheet in *freshly prepared* substrate solution. Colored bands (blue or red, depending on the enzyme and substrate used) should appear on the nitrocellulose sheet within a few minutes. When sufficient color has developed, wash the sheet in distilled water and blot dry between two sheets of filter paper.

4. Results and Discussion

This experiment is essentially a *demonstration* of the technique of protein blotting and a double-antibody method for detecting specific proteins on a protein blot. However, it should be possible, by comparing the immunologically stained paper with the stained original SDS gel, to specifically identify albumin in the acrylamide gel pattern.

References

1. Burnett, W. (1981) 'Western blotting': Electrophoretic transfer of proteins from sodium dodecyl sulfate

polyacrylamide gels to unmodified nitrocellulose. *Anal. Biochem.* **112**, 195–203.
2. Towlin, H., Stachelin, T., and Gordon, J. (1979) Electrophoretic transfer of proteins from polyacrylamide gels to nitrocellulose sheets: Procedure and some applications. *Proc. Natl. Acad. Sci. USA* **76** (9), 4350–4354.
3. Bittner, M., Kupfer, P., and Morris, C.F. (1980) Electrophoretic transfer of proteins and nucleic acids from slab gels to diazobenzyloxymethyl cellulose or nitrocellulose sheets. *Anal. Biochem.* **102**, 459–471.

Further Reading

Vaessen, R.T.M.J., Kreike, J., and Groot, G.S.P. (1981) Protein transfer to nitrocellulose filters. *FEBS Let.* **124**, 193–196.

Lin, W. and Kasamatsu, H. (1983) On the electrotransfer of polypeptides from gels to nitrocellulose membranes. *Anal. Biochem.* **128**, 302–311.

Notes on Immunoblotting. A free eight-page booklet, Mercia Brocades, Surrey, England.

Chapter 18

Thin-Layer Gel Filtration

Determination of Protein Molecular Weights

John M. Walker

1. Introduction

For many years Sephadex gel filtration of proteins has been a valuable tool for separating proteins on the basis of size difference. Sephadex is a bead-formed gel prepared by crosslinking dextran, and is available in a variety of G-types that differ in their degree of crosslinking (1). Depending on the G-type used, proteins in various molecular weight ranges can be fractionated. The basic principle of this separation method is as follows. A column is packed with Sephadex beads and buffer is pumped through to equilibrate the column. A protein mixture is then pumped onto the column, and as the sample passes through the column the individual protein components begin to separate according to their sizes. Essentially, small protein molecules, which can effectively penetrate the pores of the Sephadex beads, flow more slowly down the column than high-molecular-weight proteins that are too large to

179

penetrate *any* of the beads. Proteins of intermediate size that can enter some beads, but not others (the beads are not all of identical pore size), are correspondingly retarded to a greater or lesser extent. As a consequence, proteins elute from the bottom of the column in decreasing order of molecular weight. To apply this method to the determination of protein molecular weights, it is necessary, in a series of separate experiments, to run a number of individual proteins of known molecular weight through the column and measure the elution volume (V_e), the volume of flowing solvent required to carry the protein through the column until it emerges at maximum concentration, for each protein. A plot of V_e against log-molecular weight for each protein provides a calibration curve that is linear over a large region of the graph. The unknown protein is then run onto the column, its elution volume recorded, and the molecular weight calculated from the calibration curve. However, since each column run usually requires an overnight run, and at least six points are necessary for the calibration curve, this particular approach is rather time consuming and inconvenient. However, thin-layer gel filtration (2), which is a modification of the column method described above, allows gel filtration to be carried out on a large number of samples in only 3–4 h.

A gel of Sephadex beads (Sephadex G200 SF, fractionation range 250,000–5,000) is spread as a thin layer over a glass plate. The plate is then supported at an angle with the top of the plate connected, via a thickness of filter paper, to a buffer reservoir, which allows buffer to flow through the gel. When a steady state is achieved, aqueous solutions of protein samples (proteins of known molecular weight and an unknown) are loaded as adjacent samples along the top of the plate and left for 2–3 h to be carried through the Sephadex gel. At the end of this time the positions of the various proteins in the gel are determined by blotting the proteins out of the gel onto a sheet of filter paper and staining the paper for protein. The distance each proteins has moved in the gel is proportional to its elution volume, and hence a calibration curve of distance moved vs log-molecular weight can be constructed. Knowing the distance moved by the unknown protein, its molecular weight can be read off from the graph. This method has the advantage of allowing the samples to be compared on a single run. The total time necessary for this experiment is about 4 h, but the majority of the time involves the running of the sample on the gel plate. For teaching purposes, this experiment is therefore best run together with a second experiment that allows the time available to be used to its maximum effect, e.g., the SDS gel electrophoresis experiment, in which protein molecular weight values are also being determined. Since about 15 samples can be loaded onto each plate, it is convenient to have students in groups of two or four working on a single plate.

2. Materials Provided

A series of solutions of proteins of known molecular weight and one "unknown" sample
A microsyringe capable of delivering 5 μL
A beaker of preswollen Sephadex G-200 SF
Glass plates
Protein stain solution
Protein destain solution
Whatman 3MM and No. 1 filter paper
0.9% Saline

3. Protocol

1. Suspend the preswollen Sephadex in a beaker of 0.9% saline and allow the particles to settle for 1 h. (To save time, this step may be done by the technical staff prior to the start of the experiment.) At the end of this time, remove as much excess liquid as possible by gentle aspiration. This will give a slurry of suitable consistency for spreading.

2. Before spreading the plate, mark the underside (with a marker pen) at 0.5-cm intervals starting at 1 cm in from the edge and 3 cm from the top of the plate. This will mark the points of application for the protein samples.

3. Using the standard spreading equipment used for preparing thin-layer plates, spread the gel as a thin layer (0.2–0.9 mm) over a *clean* glass plate. The gel particles will adhere to the glass plate without added fixative.

4. Set up the thin-layer plate as shown in Fig. 1. Each plate is run at an angle of 15–20° to the horizontal. The top end of the plate is in contact with a buffer reservoir by means of a "wick" consisting of one layer of 3MM filter paper presoaked in buffer. This should be the same width as the gel layer and overlay the gel by 2 cm. A pad of moist 3MM filter paper is placed in contact with the bottom of the plate to absorb buffer as it emerges from the plate. The complete system is contained in a sealed glass tank and a saturated atmosphere maintained by placing a layer of buffer in the bottom of the tank.

   (Note: The pad of 3MM filter paper on the bottom of the gel should not be allowed to touch the buffer.)

5. Samples (*see* Table 1) *can* be loaded immediately, but it is better to allow the gel to equilibrate with the eluent for 1 h before loading. Load samples (2–3 µL) at the application points, taking care not to disturb the gel at the point of sample application.

6. Place the lid on the tank and allow the plate to run for 2.5 h.

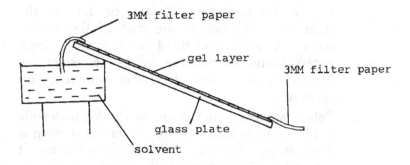

Fig. 1. Apparatus for thin-layer gel filtration.

Table 1
Some Suitable Proteins for Use in Gel Filtration on
Sephadex G-200 SF

Protein	Molecular weight
RNA polymerase (E. coli), β, and β' subunits	160,000
Aldolase	158,000
Lactate dehydrogenase	140,000
Phosphorylase b	92,500
Transferrin	78,000
Bovine serum albumin	66,300
Ovalbumin	43,000
β-Lactoglobulin	36,800
Glyceraldehyde-3-phosphate dehydrogenase	34,000
α-Chymotrypsinogen	25,700
Myoglobin	17,800
Lysozyme	14,400
Cytochrome C	12,400

7. At the end of this time, remove the glass plate from the apparatus and carefully layer a sheet of Whatman No. 1 paper over the gel, taking care to avoid trapping air bubbles, and leave for 2 min. This will allow the protein spots on the gel to transfer to the paper. Mark the corners of the gel plate on this paper.

8. Carefully peel the paper from the gel and stain for 5 min in Ponceau S. solution, followed by repeated washes in 2% acetic acid, until the background is clear. Protein samples will appear as red spots.

4. Results and Discussion

Measure the distance moved by each sample (distance from the point of origin to the *center* of the protein spot) and plot a graph of log-molecular weight vs distance moved. Measure the distance moved by the unknown sample, and determine its molecular weight from the calibration curve.

It is interesting to compare the results obtained for the same protein by both gel filtration and SDS gel electrophoresis. The results will not always be the same. The theory of molecular weight determination by Sephadex gel filtration is based on the assumption that all the standard points and the unknown protein are of the same molecular shape, which is not necessarily so. Some proteins therefore behave non-ideally on gel filtration and give anomalous results. Similarly, not all proteins fall on the calibration curve derived by SDS gel electrophoresis, particularly highly charged proteins, which bind abnormal amounts of SDS. These problems simply highlight the difficulty of obtaining an accurate molecular weight value for a give protein.

One basic difference between the gel filtration and gel electrophoresis methods is the fact that the former gives a molecular weight value of the native protein, whereas the latter will do so only if the protein is a single polypeptide chain. If the protein consists of subunits, SDS gel electrophoresis will give the molecular weights of the individual subunits.

References

1. *Gel Filtration—Theory and Practice.* A free booklet provided by Pharmacia Fine Chemicals AB, Uppsala, Sweden.
2. Andrews, P. (1971) Estimation of Molecular Size and Molecular Weight of Biological Compounds by Gel Filtration, in *Methods of Biochemical Analysis* Vol. 18, Glick, D., ed. Interscience, New York.

Chapter 19

The Dansyl-Edman Method for Peptide Sequencing

John M. Walker

1. Introduction

In 1950, Edman published a chemical method for the stepwise removal of amino acids from the N-terminus of a peptide (1). This series of reactions has come to be known as the Edman degradation, and although modifications of this technique have been introduced from time to time, the Edman degradation method remains, over thirty years after its introduction, the only effective chemical means of removing amino acids in a stepwise fashion from a polypeptide chain. The overall reaction sequence is shown in Fig. 1. The reactions may be conveniently divided into two stages:

1.1. The Coupling Reaction

In this step, the Edman reagent (phenylisothiocyanate, PITC) reacts with the amino group of the N-terminal amino acid of the polypeptide chain to give the phenylthiocarbamyl (PTC) derivative of the peptide. Amino side-chain groups of lysine residues within the polypeptide chain also couple with PITC at this stage. This is not relevant to the

185

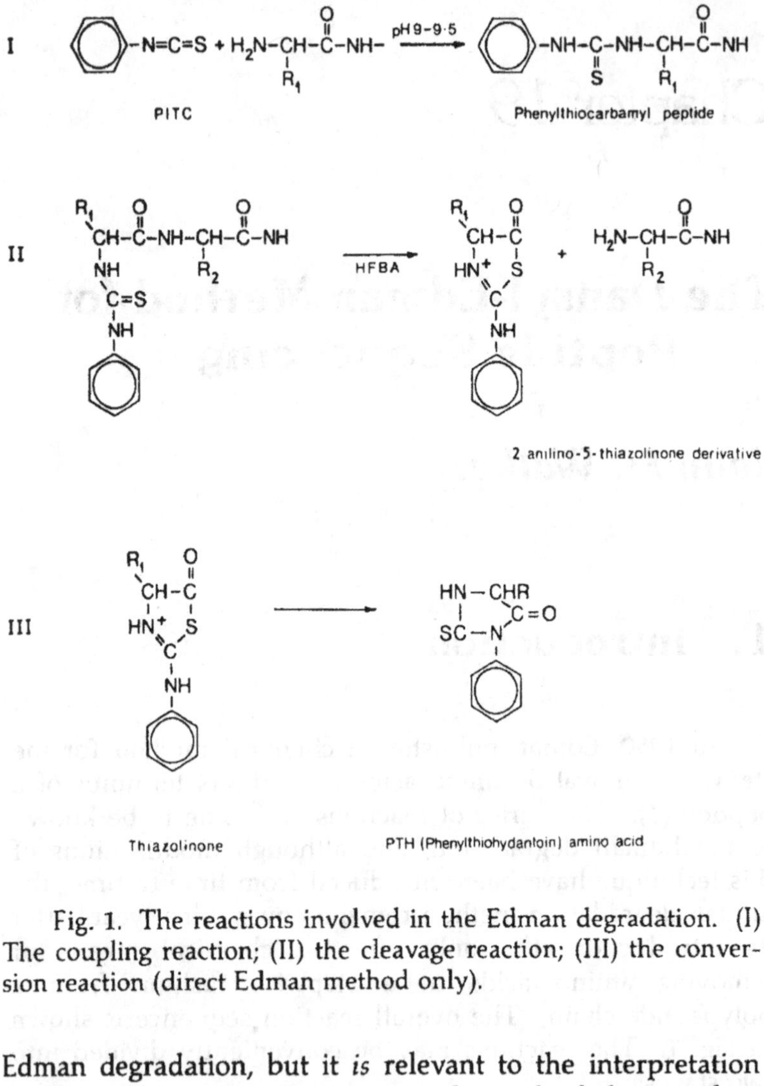

Fig. 1. The reactions involved in the Edman degradation. (I) The coupling reaction; (II) the cleavage reaction; (III) the conversion reaction (direct Edman method only).

Edman degradation, but it *is* relevant to the interpretation of the results (*see* section 4). At the end of the coupling reaction, volatile buffers and excess PITC are removed under vacuum, leaving the dried PTC derivative.

1.2. The Cleavage Reaction

In this step, the dried PTC derivative is treated with anhydrous acid. This rapid, nonhydrolytic step results in the cleavage of the PTC polypeptide at the peptide bond nearest to the PTC substituent, thus releasing the original

N-terminal amino acid as the 2-anilino-5-thiazolinone derivative, leaving the original polypeptide chain less its N-terminal amino acid. Following the cleavage reaction, the anhydrous acid is removed under vacuum and the thiazolinone derivative extracted from the remaining peptide with an organic solvent.

In the direct Edman degradation method, the extracted thiazolinone is recovered, converted to the more stable PTH derivative (conversion reaction), and identified. However, in the dansyl-Edman method (2) described here, the thiazolinone is discarded and a small aliquot of the remaining polypeptide taken for N-terminal analysis by the dansyl method. The rest of the polypeptide chain can then be subject to further cycles of the Edman degradation. After each cycle a further aliquot of the peptide is removed to determine the newly liberated N-terminal amino acid (by the dansyl method). In this way the amino acid sequence of a polypeptide chain may be determined.

The dansyl method is a highly sensitive fluorescent method for identifying amino acids. The reagent 1-dimethylaminonaphthalene-5-sulfonyl chloride (dansyl chloride) reacts with the free amino acid group of peptides according to the equations shown in Fig. 2. Total acid hydrolysis of the substituted peptide then yields a mixture of free amino acids plus the dansyl derivative of the N-terminal amino acid. The dansyl amino acid is fluorescent under UV light and identified by thin-layer chromatography, hence establishing the nature of the N-terminal amino acid of a peptide. The dansyl technique is the major method used for identifying N-terminal amino acids. The method is highly sensitive, with less than 1 nmol of the peptide being required for N-terminal analysis.

In this experiment, each student is provided with a small peptide, and carries out *one* cycle of the Edman degradation on the peptide. This then allows the first two amino acids in the peptide sequence to be identified by the dansyl method. Since an overnight hydrolysis step is required for the dansyl method, the dansyl derivatives are identified in another experiment held at a later date. The total time involved in carrying out the Edman degration and dansylating samples is about 4 h, but much of this time is "passive," involving incubation steps. The identification of dan-

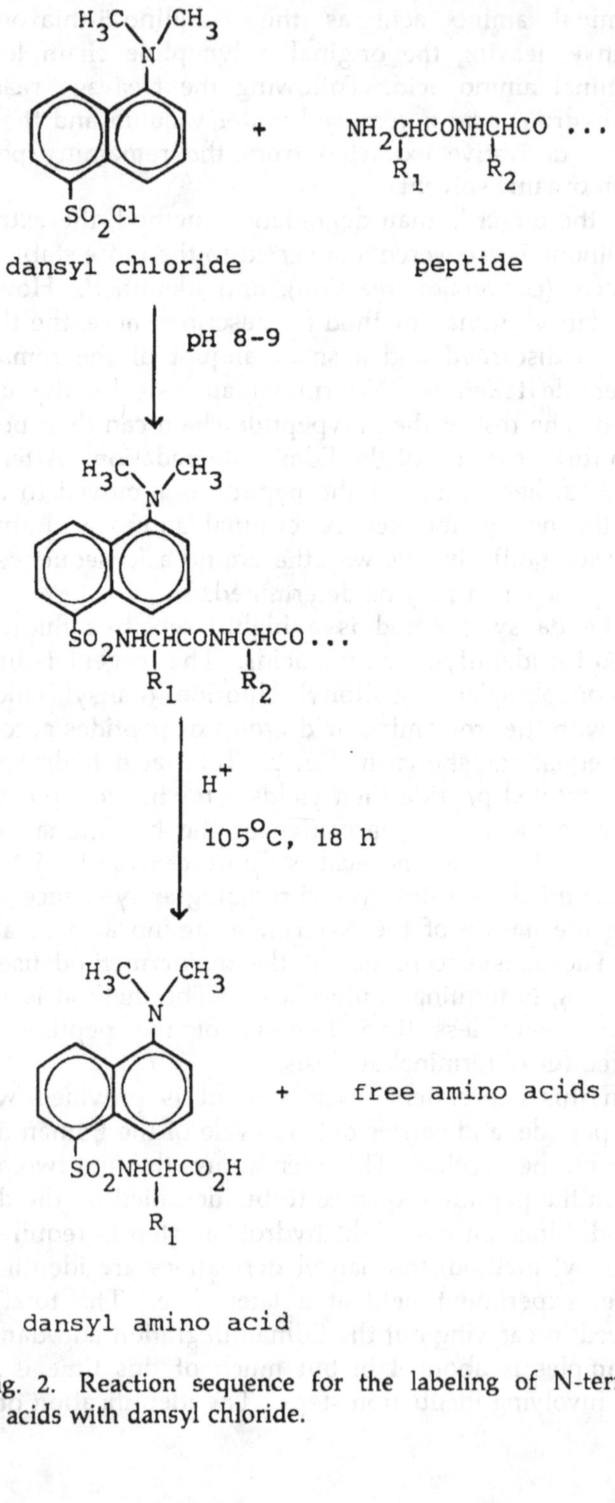

dansyl chloride   peptide

pH 8-9

H⁺

105°C, 18 h

+   free amino acids

dansyl amino acid

Fig. 2. Reaction sequence for the labeling of N-terminal amino acids with dansyl chloride.

syl amino acids, carried out at a later date, takes 2–3 h, and this period of time is fully occupied.

2. Materials Provided

2.1. The Edman Degradation

A small test tube containing an "unknown" peptide
5% (w/v) Phenylisothiocyanate in pyridine (keep in a fume cupboard)
Anhydrous trifluoroacetic acid (keep in a fume cupboard)
Water saturated n-butyl acetate
A nitrogen supply
50°C Waterbath
A vacuum desiccator (containing NaOH pellets and a beaker of P_2O_5)
Autodispensers or microsyringes

2.2. The Dansyl Method

Two "dansyl tubes"
0.2M sodium bicarbonate solution
Dansyl chloride solution
Polyamide plates, 7.5 × 7.5 cm
Chromatography tanks containing dansyl solvents 1, 2, and 3, respectively
UV lamp
A microsyringe capable of delivering 1.0 μL
37°C and 105°C ovens
Dansyl standard solution

3. Protocol

1. Dissolve the peptide in 200 μL 50% pyridine and remove a 20-μL aliquot into a dansyl tube. This 20-μL sample is labeled "A", dried under vacuum, and will be dansylated later.

2. To the rest of the peptide add 100 μL of 5% phenylisothiocyanate (PITC) in pyridine, flush the tube with N_2, and incubate at 50°C for 45 min.

3. Following this incubation, unstopper the tube and place it *in vacuo* for 30–40 min. The desiccator should contain a beaker of phosphorus pentoxide to act as a drying agent, and if possible the desiccator should be placed in a water bath at 50–60°C. When dry, a white "crust" will be seen in the bottom of the tube. This completes the coupling reaction.

4. Add 200 µL trifluoroacetic acid (TFA) to the test tube, flush with nitrogen, and incubate the stoppered tube at 50°C for 15 min.

5. Following incubation, place the test tube *in vacuo* for 5 min. TFA is a very volatile acid and evaporates rapidly. This completes the cleavage reaction.

6. "Dissolve" the contents of the tube in water (200 µL). Do not worry if the material in the tube does not all appear to dissolve. Many of the side-products produced in the previous reactions will not in fact be soluble.

7. Add *n*-butyl acetate (1.5 mL) to the tube, mix vigorously (Whirlymix) for 10 s, then centrifuge in a bench centrifuge for 3 min.

8. Taking care not to disturb the lower aqueous layer, carefully remove the upper organic layer and discard.

9. Repeat this butyl acetate extraction procedure once more, then place the test tube containing the aqueous layer *in vacuo* (with the desiccator standing in a 60°C waterbath if possible) until dry (30–40 min).

10. Redissolve the dried material in the test tube in 50% pyridine (200 µL) and remove an aliquot (20 µL) to a dansyl tube marked "B," and dry this sample under vacuum. The N-terminal amino acids of samples A and B can now be determined by the dansyl method.

11. Dissolve each sample (A and B) in sodium bicarbonate (0.2M, 10 µL), add dansyl chloride solution (10 µL), and mix.

12. Seal the tube with parafilm and incubate at 37°C for 1 h, or at room temperature for 3 h.

13. Dry the sample *in vacuo*. Because of the small volume of liquid present, this will only take about 5–10 min.

14. Add 6N HCl (50 μL) to the sample, seal the tube in an oxygen flame, and place at 105°C overnight (18 h).

15. When the tube has cooled, open the top of the tube using a glass knife, and dry the sample *in vacuo*. (This drying step is not urgent. Hydrolyzed samples may be stored at 4°C for weeks before drying.) If phosphorus pentoxide is present in the desiccator as a drying agent, and the desiccator is placed in a water bath at 50–60°C. Drying should take about 30 min. The dansylated amino acids may now be identified by thin-layer chromatography.

16. Chromatography is carried out on "polyamide" thin-layer plates (7.5 × 7.5 cm). These plates are reusable and should be placed overnight in a beaker of 1.5% formic acid after use, in order to wash off the dansyl amino acids.

    It is important that the samples are placed in the correct positions on the plates. The plates are double-sided so that a sample can be run on one side and a sample plus standards on the other. The plates carry an identifying number at the top and the point of application is marked in pencil in the bottom left-hand corner and on the reverse side.

17. Dissolve the dried sample in 50% pyridine (10 μL) and, using a microsyringe, load 1-μL aliquots at the origin on *each side* of a polyamide plate. This is best done in a stream of warm air. Do not allow the diameter of the spot to exceed 3–4 mm.

18. On the *reverse side only*, also load 0.5 μL of the standard dansyl amino acid mixture at the origin.

19. When the loaded samples are completely dry, the plate is placed in the first chromatography solvent and allowed to develop until the solvent front is about 1 cm from the top of the plate. This takes about 10 min, but can vary depending on room temperature.

20. Dry both sides of the plate by placing it in a stream of warm air. This can take 5–10 min since one is evaporating an aqueous solvent.

21. If the plate is now viewed under UV light, a blue fluorescent streak will be seen spreading up the plate from the origin; green fluorescent spots may also be seen within this streak. However, no interpretations can be made at this stage. *The viewing of dansyl plates under UV light should always be done wearing protective glasses or goggles. Failure to do so will result in a most painful conjunctivitis.*

22. Now run the dansyl plate in the second solvent, at right angles to the direction that it was run in the first solvent. The plate is therefore placed in the chromatography solvent so that the blue "streak" is along the bottom edge of the plate.

23. Run the plate in the second solvent until the solvent front is about 1 cm from the top of the plate. This takes 10–15 min.

24. Dry the plate in a stream of warm air. This will take only 2–3 min since the solvent is essentially organic. However, since benzene is involved, drying *must* be done in a fume cupboard.

25. The side of the plate containing the sample only should now be viewed under UV light. Three major fluorescent areas should be identified. Dansyl hydroxide (produced by hydrolysis of dansyl chloride) is seen as a blue fluorescent area at the bottom of the plate. Dansyl amide (produced by side reactions of dansyl chloride) has a blue-green fluorescence and is about one third of the way up the plate. These two spots will be seen on all dansyl plates and serve as useful internal markers. Occassionally, other marker spots are seen and these are described in the section 4. The third spot, which normally fluoresces green, will correspond to the dansyl derivative of the N-terminal amino acid of the peptide or protein. The separation of dansyl derivatives after solvent 2 is shown in Fig. 3. Solvent 2 essentially causes separation of the dansyl derivatives of hydropho-

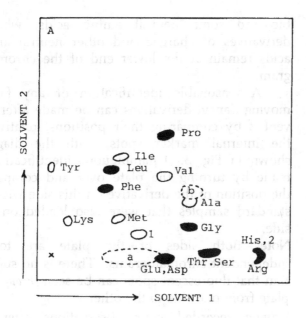

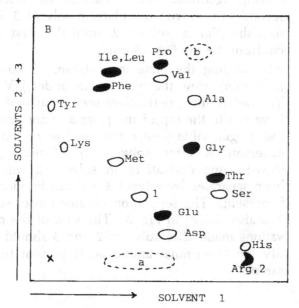

Fig. 3. Diagrams showing the separation of dansyl amino acids on polyamide plates after two solvents (A), and after three solvents (B), viewed under UV light. Abbreviations: a, dansyl hydroxide; b, dansyl amide; 1, tyrosine (o-DNS-derivative); 2, Lysine (ε-DNS-derivative). The standard dansyl amino acids that are used are indicated as black spots.

bic and some neutral amino acids, whereas
derivatives of charged and other neutral amino
acids remain at the lower end of the chromatogram.

A reasonable identification of any faster-
moving dansyl derivatives can be made after solvent 2 by comparing their positions, relative to
the internal marker spots, with the diagram
shown in Fig. 3. Unambiguous identification is
made by turning the plate over and comparing
the position of the derivative on this side with the
standard samples that were also loaded on this
side.

Note: Both sides of the plate are totally
independent chromatograms. There is no sugges-
tion that fluorescent spots can be seen *through* the
plate from one side to the other.

26. Having recorded your observations after the
second solvent, run the plate in solvent 3, in the
same direction as solvent 2, until the solvent is 1
cm from the top (10–15 min).

27. After drying the plate in a stream of warm air
(1–2 min), view the plate again under UV light.
The fast-running derivatives seen in solvent 2 will
have run to the top of the plate and are generally
indistinguishable (hence the need to record your
observations after solvent 2). However, the
slow-moving derivatives in solvent 2 will have
been separated by solvent 3 and can be identified
if present. The separation obtained after solvent
3 is also shown in Fig. 3. The sum of the observations made after solvents 2 and 3 should identify the N-terminal amino acids present in each
sample.

4.   Results and Discussion

By carrying out one cycle of the Edman degradation,
you will be able to identify the first two residues of the
peptide using the dansyl method. You should observe and

identify a single N-terminal amino acid for the first sample (indicating that the peptide is pure). This amino acid should then be completely absent in the second sample, indicating that the Edman degradation had completely removed the N-terminal amino acid. The second sample should, however, show and allow you to identify, the second amino acid in the peptide sequence.

Although the dansyl method is frequently used in conjunction with the Edman degradation as described here, it should also be stressed that it is commonly used alone as a means for determining the purity of a protein sample. A pure protein should show a single N-terminal amino acid. The presence of multiple N-terminal amino acids of varying intensity in a sample indicates an impure sample. There are a number of points concerning the dansyl-Edman method and the interpretation of the dansyl plates that need to be emphasized.

1. Manual sequencing is normally carried out on peptides between 2 and 30 residues in length and requires 1–5 nmol of peptide.

2. Since the manipulative procedures are relatively simple, it is quite normal in research work to carry out the sequencing procedure on 8 or 12 peptides at one time.

3. A repetitive yield of 90–95% is generally obtained for the dansyl-Edman degradation. Such repetitive yields usually allow the determination of sequences up to 15 residues in length, but in favorable circumstances somewhat longer sequences can be determined.

4. The sensitivity of the dansyl method is such that as little as 1–5 ng of a dansylated amino acid can be visualized on a chromatogram.

5. The side chains of both tyrosine and lysine residues also react with dansyl chloride. When these residues are present in a peptide, the chromatogram will show the additional spots o-DNS-Tyr and ε-DNS-Lys, which can be regarded as additional internal marker spots. The positions of these residues are shown in Fig. 3. These spots should not be confused with bis-DNS-Lys and

bis-DNS-Tyr, which are produced when either lysine or tyrosine is the N-terminal amino acid. For lysine residues, a strong ϵ-DNS-Lys spot will only be seen when the dansyl derivative of the N-terminal amino acid is studied. When later residues are investigated, the ϵ-DNS-Lys will be dramatically reduced in intensity or absent. This is because the amino groups on the lysine side chains are progressively blocked by reaction with phenylisothiocyanate at each coupling step of the Edman degradation.

6. The reactions of the lysine side chains with phenylisothiocyanate causes some confusion when identifying lysine residues. With a lysine residue as the N-terminal residue of the peptide, it will be identified as bis-DNS-Lys. However, lysine residues further down the chain will be identified as the α-DNS-ϵ-phenythiocarbamyl derivative because of the side chain reaction with phenylisothiocyanate. This derivative runs in the same position as DNS-phenylalanine in the second solvent, but moves to between DNS-leucine and DNS-isoleucine in the third solvent. Care must therefore be taken not to misidentify a lysine residue as a phenylalanine residue.

7. At the overnight hydrolysis step, dansyl derivatives of asparagine or glutamine are hydrolyzed to the corresponding aspartic or glutamic acid derivatives. Residues identified as DNS-Asp or DNS-Glu are therefore generally referred to as Asx or Glx, since the original nature of this residue (acid or amide) is not known. This is of little consequence if one is looking for a single N-terminal residue to confirm the purity of a peptide or protein. It does, however, cause difficulties in the dansyl-Edman method for peptide sequencing, in which the residue has to be identified unambiguously. When using the dansyl-Edman method, therefore, acid/amide residues are determined indirectly from the measurement of the peptide mobility on electrophoresis (3).

8. If the first two residues or the second and third residues in the peptide are hydrophobic, a complication will occur at the chromatography step. The peptide bond between two such residues is particularly (although not totally) resistant to acid hydrolysis. Under normal conditions, therefore, some dansyl derivative of the first amino acid is produced, together with some dansyl derivative of the N-terminal dipeptide. Such dipeptide derivatives generally run on chromatograms in the region of phenylalanine and valine. However, their behavior in solvents 2 and 3, and their positions relative to the marker derivatives, should prevent mis-identification as phenylalanine or valine. Such dipeptide spots are also produced when the first residue is hydrophobic and the second residue is proline, and these dipeptide derivatives run in the region of proline. However, since some of the N-terminal derivative is always produced, there is no problem in identifying the N-terminal residue when this situation arises.

References

1. Edman, P. (1956) On the mechanism of the phenyl-isothiocyanate degradation of peptides. *Acta Chemica Scand.* **10**, 761.
2. Hartley, B.S. (1970) Strategy and tactics in protein chemistry. *Biochem. J.* **119**, 805.
3. Offord, R.E. (1966) Electrophoretic mobilities of peptides on paper and their use in the determination of amide groups. *Nature* **211**, 591–593.

Further Reading

Protein Sequence Determination. (Needleman, S.B., ed.) (1970) Springer-Verlag, New York.

Advanced Methods in Protein Sequence Determination. (Needleman, S.B., ed) (1977) Springer-Verlag, New York.

Introduction to Protein Sequence Analysis. L.R. Croft (1980) John Wiley & Sons. Chichester, England.

Methods in Peptide and Protein Sequence Analysis. (Birr, C., Ed.) (1980) Elsevier, North Holland, Amsterdam.

Techniques in Molecular Biology. (Walker, J.M. and Gaastra, W., eds.) (1983) Croom Helm, Beckenham, Kent, England.

Chapter 20

The Double Immunodiffusion Technique

Immunoprecipitation and Analysis of Antigenic Protein in Gel

David L. Cohen

1. Introduction

Antibodies are globulin proteins produced by vertebrates in response to specific antigenic (immunogenic) challenge, and are a major feature of vertebrate immune defense against microbial disease. Antigens (immunogens) are substances capable of eliciting such responses because of their innate foreignness to the exposed animal. The nature of this foreignness relates to the presence of discrete chemical groupings (antigenic determinants or epitopes) on the surface of the antigenic molecule. All antigenic molecules consist of a mosaic of similar or dissimilar epitopes, with the number of epitopes representing the valency of the antigen. An important attribute of the antibody molecules is that they can react specifically with an antigen because they possess binding sites that are complementary to the antigenic epitopes. A good analogy is used by Feinberg (1), who refers to the epitope "bump" fitting the antibody "hollow." Whereas the antigen is almost certainly multivalent, the majority of globulin antibody proteins in serum are divalent (IgG), i.e., possess two linked identical binding sites. What consequences spring from these characteristics? When a specific antiserum is mixed in vitro with the corresponding antigen, if the conditions are correct, i.e., pH, temperature, and correct balance of electrolytes, the antibody combines reversibly with antigen to form an aggregate, precipitate, or floccule. This interaction between specific antibody and antigen can be used as the basis of analytical techniques to identify and quantify protein.

The single most important factor in immunoprecipitation, apart from the obvious one of specificity of the individual reactants, is that of the ratio of the reactants. The optimal ratio will lead to the shortest precipitation time. In excess of either reactants, either inhibition of immunoprecipitation occurs or soluble complexes are formed. In fact, there is a narrow range of relative concentrations of antibody and antigen in which there is no free antigen or antibody present after the formation of the immunoprecipitate—this is called "the equivalence zone." This zone occurs usually before the point of maximum precipitation and thereafter the increased amounts of antigen are incorporated into the precipitate, which increases in size. As the antigen concentration increases, further soluble complexes are formed that lead to the decrease in the amount of immunoprecipitation (Figs. 1 and 2).

Immunoprecipitation in a liquid phase, as described above, can be very specific, but it is not very discriminating when dealing with more than one reaction since it is possible, for example, for two separate immunoprecipitates to form if the antiserum contains antibodies against two specific, but differing, antigens. Thus, there have been devised a number of improvements to this technique of immunoprecipitation, all of which depend upon the passive diffusion of antigen and antibody toward each other in a gel, usually agar or agarose.

In the technique devised by Oudin and adapted by Oakley and Fulthorpe (2), a layer of molten agar containing antiserum is placed in a test tube and allowed to set.

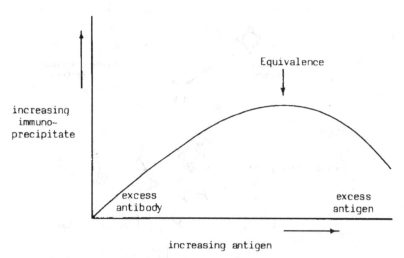

Fig. 1. *Precipitation curve of antigen with corresponding antibody.*

Increasing amounts of antigen are added to a fixed amount of antibody. The supernatant fluid is assayed for free antibody or antigen, while the quantity of immunoprecipitate can be determined spectrophotometrically or turbidometrically. The point or zone where no free antibody or antigen can be detected is called "the zone of equivalence."

Another layer of agar containing the antigen is then allowed to set above this. If necessary, a middle layer of agar can be used to separate the reactants. Passive diffusion of reactants occurs and disks of immunoprecipitates form after about 24–48 h. The advantage of this system is that a concentration gradient is set up within the gel of both reactants; where they meet in optimal concentrations, an immunoprecipitate is formed. This is one example of what is called *double immunodiffusion*. Another variation of this theme is the plate technique devised by Ouchterlony in 1958. This is an elegant adaptation of the gel diffusion technique and is quite discriminating and easy to set up. Since its early days, it has itself been improved, adapted, and modified by Ouchterlony and others, and the reader is referred to the excellent texts by Weir (3) and Axelson (4) for more details.

Despite the advent of new immunological techniques, immunoprecipitation by double immunodiffusion continues to be a useful laboratory tool for the identification and comparison of precipitable antigenic material, not least

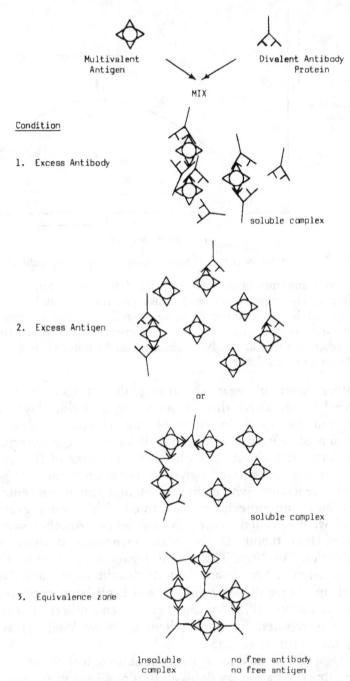

Fig. 2. Diagrammatic representation of types of immunoprecipitates formed by union of a multivalent antigen with a divalent
antibody (IgG) in the precipitation curve experiment illustrated in
Fig. 1.

because of its technical simplicity and high resolving power.

Four experiments are outlined below that illustrate some of the possible interactions that can occur between antigen and antibody when they diffuse toward each other from their respective origins in wells cut in a thin layer of agarose gel on the surface of a microscope slide.

Experiment 1 illustrates the specificity of an antiserum (anti-bovine serum albumin) by its ability to precipitate only bovine serum albumin and not human serum albumin or ovalbumin. Experiment 2 illustrates the ability of two specific antisera (anti-bovine serum albumin and anti-bovine γ-globulin) to cause the separate precipitation of bovine serum albumin and bovine γ-globulin. The nature of the individual precipitation bands illustrates that the proteins that have been precipitated are antigenically dissimilar. Experiment 3 illustrates two points: (1) that an antibody to normal calf serum contains antibodies to all the protein components of the serum and thus produces multiple precipitation bands, and (2) that the anti-calf serum will not precipitate proteins in horse, human, or rabbit sera. Experiment 4 is an illustration of a "forensic-type" experiment to identify an unknown serum "x" (in this case using the standard anti-species sera).

Each experiment, provided the molten agarose gel, antiserum, and antigen solutions have been prepared, can be set up in under a half-hour. The slides, once prepared, are placed in a humid box and incubated at 37°C for 6–24 h. The precipitation patterns can then be directly ascertained by using a box with oblique lighting or by the more lengthy process of staining, which would take approximately one more hour.

2. Materials Provided

2.1. Day 1

Molten 1% agarose in phosphate buffered saline
Clean microscope slides
Leveling table
Sterile Pasteur, 5- and 10-mL pipets
Small rubber teats

Hypodermic syringes and needles
Humid box
Filter paper
Plastic sandwich box with lid

2.1.1. Reagents for Expt. 1

Bovine serum albumin at 2.5 and 5.0 mg/mL
Human serum albumin or ovalbumin at 5.0 mg/mL
Anti-bovine serum

2.1.2. Reagents for Expt. 2

Calf serum at 1/10
Anti-bovine serum albumin
Anti-bovine γ-globulin

2.1.3. Reagents for Expt. 3

Calf serum at 1/10, 1/100, 1/1000
Horse, rabbit, and human serum at 1/10
Anti-calf serum

2.1.4. Reagents for Expt. 4

Unknown serum x at 1/10
Anti-calf serum
Anti-horse serum
Anti-human serum

2.2. Day 2

Light box
Plastic sandwich box for staining/destaining
Coomassie blue stain
Destaining solution

3. Protocol

3.1. Day 1

1. Using a sterile 5- or 10-mL pipet, gently and slowly add 5 mL 1% agarose solution to the surface of a clean microscope slide placed on the lev-

eling table. Be careful not to let the solution run off the edge of the slide.

2. When the agarose has set, the slide can be placed for a few minutes in a refrigerator at 4°C to make the gel a little harder, enabling the next stage of gel-cutting to be performed with ease.

3. Place the slide over one of the four templates provided below.

3.1.1. Expt. 1. Demonstration of Identity

The following template can be used:

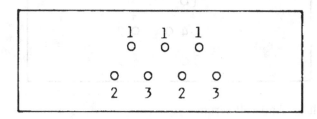

Upper row: wells 1 Bovine serum albumin 5 mg/mL
(antigens)   wells 2 Bovine serum albumin 2.5 mg/mL
              wells 3 Human serum albumin or
              Ovalbumin 5 mg/mL
Lower row: wells 4 Anti-bovine serum albumin
(antiserum)

3.1.2. Expt. 2. Demonstration of Identity and Nonidentity

The following template can be used:

Upper row: wells 1 Calf serum 1/10
(antigen)

Lower row: wells 2 Anti-bovine serum albumin
(antisera)   wells 3 Anti-bovine γ-globulin

3.1.3. Expt. 3. Demonstration of Identity and Multiple Precipitation Bands

The following template can be used:

```
        2 o
    1 o        o 3
        o
    5 o  7  o 4
      6 o
```

well 1   Calf serum 1/10
well 2   Calf serum 1/100
well 3   Calf serum 1/1000
well 4   Horse serum 1/10
well 5   Rabbit serum 1/10
well 6   Human serum 1/10
well 7   Anti-calf serum

3.1.4. Expt. 4. Identification of an Unknown Serum

The following template can be used:

```
    1 o

      4 o   o 2

    3 o
```

well 1   Anti-calf serum
well 2   Anti-horse serum
well 3   Anti-human serum
well 4   Unknown serum x 1/10

4. Holding a Pasteur pipet erect and the teat bulb compressed, cut a well in the desired position in the agarose, releasing the pressure on the teat bulb. This will enable the plug of agarose to be drawn into the Pasteur pipet. (A little practice may be required.) Repeat the gel-cutting procedure in the desired pattern on the template.

5. Fill the wells with solutions of antigens or antiserum until the meniscus in the well disappears. The filling is performed with labeled injection syringes (1.0 mL) with no. 25G needles attached. (Overflows can be mopped up quickly with small pieces of filter paper.)
   Note: *Be careful not to contaminate the reagents, i.e., keep each labeled syringe next to the specific reagent.*

6. Place the slide, when filling is complete, into a humid chamber and incubate at 37°C for 6–24 h, or leave at room temperature for 24 h.

3.2. Day 2

1. Examine the slides for precipitation lines. This is best performed using a light box that gives oblique lighting to the slide, on a dark background. If a permanent record is desired, the slide can be stained using Coomassie Brilliant blue and destained with an acetic acid/methanol/water destaining solution, as described below.

2. It is possible to speed the process of staining and destaining if the slide is first dried to a thin film of agarose. This is performed by placing a folded pad of Whatman No. 1 filter paper over the slide (agarose side up) and placing a heavy weight on top of this pad of paper.

3. After 0.5–1 h, the filter paper must be dampened and can then be carefully removed. The agarose layer can be further dried using a hair dryer, if necessary.

4. The thin dry film of agarose can now easily be subjected to the process of staining. Place the

dried slide, agarose side uppermost, into the sandwich box, and cover with a layer of Coomassie Brilliant blue stain and leave for 5–10 min.

5. Remove the slide. Dispose of the stain. Replace the slide in the sandwich box and cover with a layer of the destaining solution. Gently agitate the box for up to 1 h until fully differentiated, i.e., when the background is colorless and the precipitation lines remain stained. It is possible to change the destaining solution during the process of differentiation.

Note: The unprecipitated protein can be removed from the gel before commencing the drying and staining procedures. This is performed by giving the slides several washes in phosphate buffered saline in a shallow dish (being careful not to wash the agarose off the slide).

4. Results and Discussion

4.1. General Interpretation of Precipitation Patterns

Many precipitation patterns are possible between antigens and antibodies; the following diagrams and interpretations A to E illustrate a range of likely patterns obtainable in gel diffusion experiments.

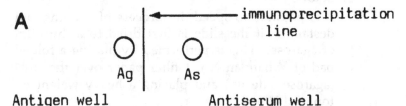

A

immunoprecipitation line

Ag   As

Antigen well   Antiserum well

Interpretation: In A, it can be seen that there is a line of immunoprecipitation, indicating that the antiserum has precipitated one single antigen, i.e., only one antigen/antibody reaction took place that produced a visible immunoprecipitation.

B

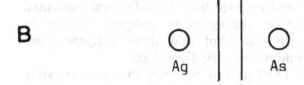

Ag | As

Interpretation: In B, there are two separate lines of immunoprecipitation separating the wells of
reagents, indicating that the antigen well
contains two discrete antigens that could be
separately precipitated by the antiserum.
(Conversely, one might say that the
antiserum contains specific antibody to two
distinct antigens present in the antigen well.)

C

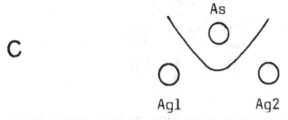

As

Ag1 Ag2

Interpretation: In C, two antigen wells are filled with
respective antigens 1 and 2. This would
indicate that the antiserum contained
specific antibodies to antigen present in both
antigen wells, and that this antigen was the
same in both wells, i.e., the antiserum has
indicated *immunological identity* between the
antigens present in Ag1 and Ag2.

D

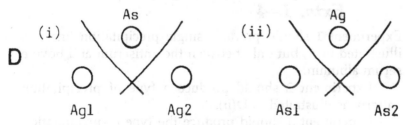

(i) As

Ag1 Ag2

(ii) Ag

As1 As2

Interpretation: In D(i), it can be seen that two separate lines
or arcs of immunoprecipitation have been
formed by reaction between the antiserum
and antigen present in Ag1 and Ag2. This

would indicate that the antiserum contained specific antibodies to antigens present in both wells, but that these antigens were immunologically dissimilar.

In D(ii), it can be seen that two separate lines or arcs of immunoprecipitation have been formed by reaction between the two specific antisera (As1 and As2) and the antigen present in Ag. This would indicate that the antigen well contained two specific and immunologically unrelated antigens, i.e., the antigen was not homogeneous.

E

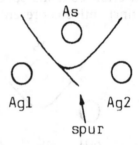

Interpretation: In E, the precipitation pattern would indicate that the antiserum contained antibodies to both antigens 1 and 2, but that these antigens were only partially identical (showed common determinants, epitopes, present), since a spur has formed at one end of an arc of precipitation.

4.2. *Likely Results of Gel Diffusion Exts. 1–4*

Experiment 1 should produce single precipitation lines, as illustrated in A, but only between the antiserum and bovine serum albumin.

Experiment 2 should produce a type of precipitation pattern, as illustrated in D(ii).

Experiment 3 should produce the type of precipitation pattern illustrated in B and C. Multiple arcs of immunoprecipitate between anti-calf serum and calf serum would indicate that the anti-calf serum contains specific antibodies to all of the major antigenic components of calf serum.

In Expt. 4, depending upon the nature of the unknown serum x, there should, of course, only be a precipitation reaction between the unknown serum and one of the specific antisera.

References

1. Feinberg, G. and Jackson, M.A. (1984) *The Chain of Immunology*, Blackwell Scientific Publications, Oxford.
2. Oakley, C.L. and Fulthorpe, A.J. (1953) Antigenic Analysis by Diffusion. *J. Pathol. Bacteriol.* **65,** 49–60.
3. Weir, D.M. (ed.) (1978) *Handbook of Experimental Immunology* *Vol. 1.* Blackwell Scientific Publications, Oxford.
4. Axelson, N.H. (ed.) (1983) *Handbook of Immunoprecipitation in Gel Techniques. Scand. J. Immunol.* (Suppl.) **10, 17.**

Chapter 21

Immunoelectrophoretic Techniques

2-D Immunoelectrophoresis and Rocket Immunoelectrophoresis

John M. Walker

1. Introduction

This experiment introduces the student to two commonly used immunoelectrophoretic techniques; rocket immunoelectrophoresis, also known as electroimmunoassay, and 2-D immunoelectrophoresis.

Rocket immunoelectrophoresis is a simple, quick, and reproducible method for determining the concentration of a specific protein in a protein mixture. The method, originally introduced by Laurell (1), involves a comparison of the sample of unknown concentration with a series of dilutions of a known concentration of the protein, and requires a monospecific antiserum against the protein under investigation. The samples to be compared are loaded side by side in small circular wells along the edge of an agarose gel that contains the monospecific antibody. These samples

213

(antigen) are then electrophoresed into the agarose gel, where interaction between antigen and antibody takes place. In the presence of excess antigen, the antigen-antibody complex is soluble, but as the antigen moves further into the gel, more antigen combines with antibody until a point of equivalence is reached (*see* Chapter 20). At this stage, the antigen-antibody complex is insoluble. The end result is a precipitation "rocket" spreading out from the loading well. Since the height of the peak depends on the relative excess of antigen over antibody, a comparison of the peak heights of the unknown and standard samples allows the unknown protein concentration to be determined. Although particularly useful for measuring the concentration of any given protein in a serum sample, the method may also be used to measure, or compare, protein concentrations in any mixture (e.g., urine, cerebrospinal fluid, tissue homogenates, and so on) as long as an antiserum to the protein is available.

2-D immunoelectrophoresis consists of two sequential electrophoretic stages:

(i)   The first dimension, during which the protein mixture (usually serum proteins) is separated by simple electrophoresis in an agarose gel.

(ii) The second dimension, during which the separated proteins are electrophoresed at right angles into a freshly applied layer of agarose containing a predetermined amount of antibody.

During the second electrophoretic step, the proteins migrate into the gel until they are completely precipitated by the antiserum, each protein forming a separate precipitate peak. The area under any peak is directly proportional to the concentration of that protein in the serum and inversely proportional to the concentration of specific antibody in the gel.

This technique is particularly useful for quantitatively measuring changes in serum protein patterns in human disease. As many as 40 different serum proteins may be quantitated in a single experiment.

In the experiments presented here, rocket electrophoresis is carried out using a series of dilutions of bovine serum albumin that provide data for a calibration graph. A dilute bovine serum sample is run at the same time, and from the measurements of the peak height of this sample and by reference to the calibration graph, the concentration of albumin in the original serum sample can be determined.

For the two-dimensional electrophoresis, a bovine serum sample can be studied, the second dimension involving electrophoresis into a gel containing anti-bovine serum. However, an alternative, and more interesting, approach is to use the student's own serum, the second dimension involving electrophoresis into a gel containing anti-whole human serum. Since such small volumes are required, the serum sample is easily obtained by finger puncture. All the necessary antisera are commercially available, and since such small volumes are used, the experiment is relatively inexpensive.

The size of gel described in this experiment (5 × 5 cm) for rocket immunoelectrophoresis is suitable for practical demonstrations. However, it should be pointed out that for highly accurate determinations, larger plates should be used (7.5 × 7.5 cm) with larger wells (5–10 μL volumes) so that sample loadings can be made with greater accuracy, and the correspondingly larger peaks measured more accurately. When possible, for both experiments the gel plates used should be as thin as possible to maximize the effect of the cooling plate. This is particularly important in the case of the shorter run-time when considerable heat is evolved, which can cause distortion of the precipitation peaks. Because of the thickness of the electrophoresis wicks used, they are quite heavy when wet and have a tendency to "slip off" the gel during the course of the run. This can be prevented by placing thick glass blocks on top of the wicks where they join the gel, or by placing a heavy glass sheet across both wicks, which, by also covering the gel, has the added advantage of reducing evaporation from the gel. A uniform field strength over the entire gel is critical in rocket immunoelectrophoresis. For this reason the wick should be exactly the width of the gel. When setting up the electrophoresis wicks, it is important that they be kept well away from the electrodes. Since the buffer used in both experiments is of low ionic strength, if electrode products are allowed to diffuse into the gel, they can ruin the gel run. Most commercially available apparatus are therefore

designed such that the platinum electrodes are masked
from the wicks and also separated by a baffle. Most commercially available electrophoresis apparatus will hold four
gel plates. The organization of the class is generally limited
by the number of electrophoresis apparatus available, but if
possible, students should work in pairs, each pair producing one rocket gel and one 2-D gel.

   The first dimension for crossed immunoelectrophoresis
takes about 1½ h to set up and run: the second dimension
takes about 10 min to set up, and is then run overnight.
Rocket immunoelectrophoresis takes about 30 min to set up
and is best run overnight, although if an efficient cooling
plate is available, it can be run at a higher current for 2½ h.
Following electrophoresis, both gels must be dried and
stained for protein. However, it is not essential to carry out
these steps immediately after electrophoresis, since the gels
may be stored in 0.14M saline until a later date. When
carried out, the drying and staining of gels can be achieved
in less than 1 h.

2. Materials Provided

2.1. Rocket Immunoelectrophoresis

Rabbit anti-bovine serum albumin antibody
Dilutions of a stock solution of bovine serum albumin
A known dilution of bovine serum

2.2. 2-D Immunoelectrophoresis

Rabbit anti-bovine serum or anti-human serum antibody
Marker bovine serum containing bromophenol blue
Bovine serum (not necessary if students are analyzing their
   own serum)

2.3. General

0.06M Barbitone buffer, pH 8.4
2% Aqueous agarose in a waterbath at 52°C
Clean glass plates, 5 × 5 cm
Protein stain

Destain solution
Strips of Whatman No. 1 filter paper, 5 cm wide
A flat-bed electrophoresis apparatus with cooling plate
Microsyringes

3. Protocol

3.1. Rocket Immunoelectrophoresis

1. Stand a 5-mL pipet in the agarose solution and allow time for the pipet to warm up.

2. Place a test tube in the 52°C waterbath and add 2.8 mL barbitone buffer (0.06M). Leave this for about 3 min to allow the buffer to warm up, then add agarose solution (2.8 mL). This transfer should be made as quickly as possible to avoid the agarose setting in the pipet. Using a pipet with the end cut off to give a larger orifice allows for a more rapid transfer of the viscous agarose solution. Some setting of agarose in the pipet is inevitable, but is of no consequence. Briefly mix the contents of the tube and return to the waterbath.

3. Thoroughly clean a 5 × 5 cm glass plate with methylated spirit. When dry, put the glass plate on a leveling plate or level surface.

4. Add rabbit anti-bovine serum albumin (50 µL) to the diluted agarose solution and *briefly* mix to ensure even dispersion of the antiserum. Briefly return to the waterbath to allow any bubbles to settle out.

5. Immediately pour the contents of the tube onto the glass plate so that it totally covers the surface of the plate. Keep the neck of the tube close to the center of the plate and pour slowly. Surface tension will prevent the liquid from running off the edge of the plate. Alternatively, tape can be used to form an edging to the plate. The final gel will be approximately 2 mm thick.

6. Allow the gel to set for 10 min, then make holes (~1 mm diameter) at 0.5-cm spacings, 1 cm from

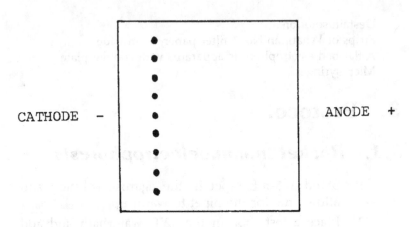

Fig. 1. Diagram showing the positioning of loading wells for a rocket immunoelectrophoresis plate.

one edge of the plate. This is most easily done by placing the gel plate over a predrawn (dark ink) template (such as the one shown in Fig. 1), when well positions can easily be seen through the gel. The wells can be made using a Pasteur pipet or a piece of metal tubing attached to a weak vacuum source (e.g., a water pump).

7. Pour approximately 750 mL of 0.03M barbitone buffer into each reservoir of the apparatus and completely wet the wicks (eight layers of Whatman No. 1 paper) with buffer. Place the gel plate on the cooling plate of an electrophoresis tank and place the electrode wicks over the edges of the gel. Take care not to overlap the sample wells and ensure that these wells are nearest the cathode.

8. Deliver 1 μL of each of the BSA dilutions into separate wells. Place 1 μL of the diluted serum sample into one well.

9. Place the wicks on the edges of the plate and hold the wicks in place with glass blocks.

10. Immediately pass a current of 20 mA ($\sim$200 V) for 2.5 h. (This is $\sim$10 V/cm across the plate.) Alternatively, run at 2.5 mA overnight.

Note: This current applies for one gel. Increase the current accordingly if more than one gel is to be run on the apparatus.

11. At the end of the run, precipitation rockets can be seen in the gel. These are not always easy to visualize and are best observed using oblique illumination of the gel over a dark background.

12. A more clear result can be obtained by staining these precipitation peaks with a protein stain. First the antiserum in the gel, which would otherwise stain strongly for protein, must be removed. One way of doing this is to wash the gel with numerous changes of saline (0.14M) over a period of 2–3 d. The washed gel may then be stained directly, or dried and then stained, to give a permanent record. To dry the washed gel, place it on a clean sheet of glass, wrap a piece of wet filter paper around the gel, and place it in a stream of warm air for about 1 h. Wet the paper again with distilled water and remove it to reveal the dried gel fixed to the glass plate. Any drops of water on the gel can be finally removed in a stream of warm air.

13. A more convenient and quicker method of preparation is to blot the gel dry. With the gel on a sheet of clean glass, place eight sheets of filter paper (Whatman No. 1) on top of the gel and apply a heavy weight (lead brick) for $\frac{1}{2}$ h. After this time, carefully remove the now wet filter papers to reveal a flattened and nearly dry gel. Complete the drying process by heating the gel in a stream of hot air for about 1 min. When completely dry, the gel will appear as a glassy film.

14. Whichever method of gel preparation is used, the gel can now be stained by placing in stain (with shaking) for 10 min, followed by washing in destain.

15. Peak heights can now be measured and a graph of concentration against peak height plotted, and the albumin concentration in the original serum sample determined. A typical electrophoretic run is shown in Fig. 2.

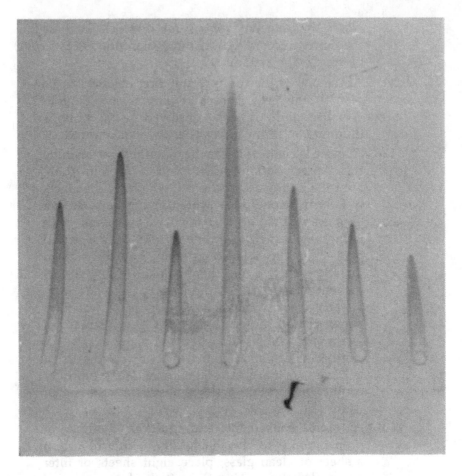

Fig. 2. A rocket immunoelectrophoresis gel (5 × 5 cm) run
at 20 mA for 3 h and stained for protein with Coomassie Brilliant
Blue. The gel contained rabbit anti-bovine serum albumin, and
the sample loadings (1 μL/well) were the following dilutions of a
stock 4% BSA solution (left to right):

   1:200
   1:150
   A 1:300 dilution of bovine serum
   1:100
   1:200
   1:300
   1:500

3.2. 2-D Immunoelectrophoresis

1. Place a test tube in the waterbath and add 2.8 mL 0.06M barbitone buffer to the test tube.

2. Leave for 3 min, then add 2.8 mL of agarose solution. This transfer should be made as quickly as possible to avoid the agarose setting in the pipet. Briefly Whirlymix the contents and return to the waterbath.

3. Clean a 5 × 5 cm glass plate thoroughly with methylated spirit. Dry and place on the leveling plate.

4. Remove the test tube from the waterbath and pour the contents of the tube onto the glass plate. Keep the neck of the tube close to the plate and pour slowly. Surface tension will keep the liquid on the surface of the plate. Alternatively, tape can be used to form an edging to the plate.

5. Allow the gel to set for 5 min. While the gel is setting, pour 800 mL 0.03M barbitone buffer into each reservoir of the electrophoresis tanks and completely wet the wicks.

6. Using a Pasteur pipet, make two holes in the gel 0.8 cm in from one edge of the plate and approximately 1.5 cm in from the side of plate (see Fig. 3).

7. Place the plate on the cooling plate of the electrophoresis apparatus. Place 0.5 μL of serum (human or bovine) in the test well and 0.5 μL of reference serum in the other well. "Top-up" the wells with electrode buffer ($\sim$1 μL). (Human serum samples can be obtained by pricking the thumb and centrifuging 1–2 drops of blood in a 1.5-mL Eppendorf tube for 10 s in a microfuge. The straw-colored serum can then be easily removed with a microsyringe.)

8. Place the wicks on the edges of the plate (almost touching the wells on the sample side) and hold the wicks in place with glass blocks.

9. Immediately pass a current of 20 mA ($\sim$250 V) until the blue marker dye (which is bound to the albumin) in the reference sample reaches or just

First dimension electrophoresis

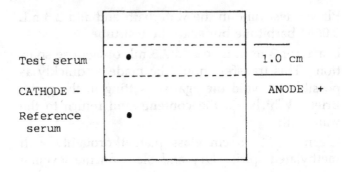

After electrophoresis the gel is cut where indicated by
the dotted line.

Second dimension electrophoresis

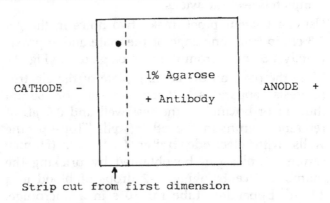

Fig. 3. Diagram showing the positioning of loading wells for
a 2-D immunoelectrophoresis plate.

enters the anode wick. This will take approximately 40 min. (This is about 10 V/cm across the
plate.)

10. While this plate is running, repeat steps 2 and 3, using a new test tube, but this time use only 2.2 mL of buffer and agarose.

11. When the plate has run, return the plate to the leveling table and cut the gel as shown in Fig. 3. Transfer the gel slice to a second clean glass plate.

12. To the tube in the waterbath now add 200 μL of rabbit anti-bovine serum antibody (or anti-human serum antibody) and briefly Whirlymix to ensure even dispersion of antibody. Return briefly to the waterbath to allow any bubbles to settle out.

13. Pour the agarose and antibody mixture onto the second plate adjacent to the strip of gel and allow the gel to set for 5 min.

14. Place the plate in the electrophoresis apparatus as before and run at 2.5 mA/plate overnight (1 V/cm).

15. Following the overnight run, precipitation peaks will be seen in the gels. These peaks can be enhanced by drying the gel and staining for protein as described for rocket immunoelectrophoresis. A typical 2-D run on human serum is shown in Fig. 4.

4. Results and Discussion

4.1. Rocket Immunoelectrophoresis

Following staining and destaining, measure the height (mm) of each rocket and plot a graph of peak height vs concentration of BSA. You should get a good straight-line graph. Measure the peak height of your unknown sample and, using your graph, determine the concentration of BSA in this sample.

4.2. 2-D Immunoelectrophoresis

The 2-D result is basically visual, but individual protein can be identified by comparison with a suitable map (e.g., *see* refs. 2 and 3).

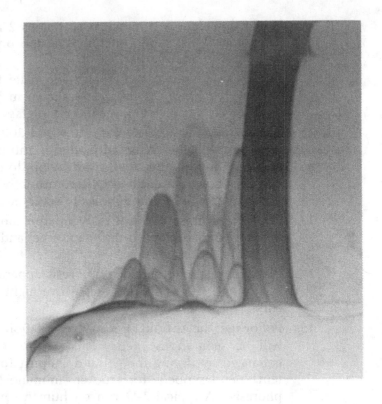

1st dimension

Fig. 4. A typical 2-D immunoelectrophoresis pattern of human serum, stained with Coomassie Brilliant Blue. Some of the weaker immunoprecipitation peaks are not visible on this photograph.

References

1. Laurell, C-B. (1966) Quantitative estimation of proteins by electrophoresis in agarose gel containing antibodies. *Anal. Biochem.* **15**, 45–52.

2. Minchin Clarke, H.G. and Freeman, T. (1968) Quantitative immunoelectrophoresis of human serum proteins. *Clin. Sci.* **35**, 403–413.
3. Dako Information Sheet No. 4. Provided by DAKO-immunoglobulins a/s, Copenhagen, Denmark (or local supplier).

Further Reading

A Manual of Quantitative Immunoelectrophoresis—Methods and Applications (1973) *Scand. J. Immunol.* **2**, (Suppl. 1) (Axelsen, N.H., Kroll, J., and Week, B., eds.) Universitatsforgleget, Oslo.

Handbook of Immunoprecipitation-in-Gel Techniques (1983) (Axelsen, N.H., Ed.) Blackwell Scientific Publications.

Chapter 22

The Time Course of β-Galactosidase Induction in *Escherichia coli*

Kenneth H. Goulding

1. Introduction

Bacteria can vary their enzymic composition markedly when grown in different conditions. In particular, specific enzymes required to utilize a given carbon or nitrogen source are usually only present when they are essential for growth. Such enzymes are known as *inducible*. In contrast, some enzymes that are normally present in actively growing cells may under certain conditions, especially if the product of their reaction(s) is present in excess in the growth media, be absent or present only in barely detectable amounts. Such enzymes are known as *repressible*. Naturally certain enzymes are present in cells, in variable amounts, regardless of growth conditions. These enzymes are known as *constitutive*. This ability to regulate the levels of enzymes in response to different growth conditions enables bacteria to maximize their resourses by only synthesizing enzymes when they are *actually needed* to support cell maintenance and growth.

227

The mechanisms of enzyme induction and repression are a form of control of gene expression that has been studied in great detail in bacterial systems. Indeed, an examination of any modern cell biology, biochemistry, or genetics text book (1–4) will indicate the detailed current knowledge of how control of these processes is achieved. It is not the purpose of this article to discuss these mechanisms, and the reader is referred to one of the books cited above.

In the experiment described in this chapter, the time course of induction of β-galactosidase in the enteric bacterium *Escherichia coli* is examined. This enzyme is required when *E. coli* utilizes the disaccharide lactose as a growth substrate so that the substrate can be hydrolized to its two component monosaccharides, glucose and galactose, that are then further metabolized by constitutive enzyme systems. A strain of *E. coli* that grows very slowly using glycerol as the sole carbon substrate is used. When lactose is added to cultures growing on glycerol it becomes the preferred substrate, and β-galactosidase is rapidly induced. An interesting feature of the experiment is the comparison of the time course of induction in response to isopropylthiogalactoside (IPTG) or lactose. IPTG induces β-galactosidase because of its structural similarity to lactose, but it is not able to act as a growth substrate because it is not hydrolyzed by the enzyme. It is therefore known as a *gratuitous inducer*.

The experiment described is suitable for a practical class of 5 or 6 h duration. Using the basic techniques described, which are relatively straightforward and do not require elaborate facilities or equipment, a whole series of further experiments are possible, some of which are suggested in the discussion. These could extend the practical work to one of several 6-h sessions or would be suitable for a small project. It is worth comment at this stage that the experiment described is only suited to students working in pairs or threes, because at certain times several things need to be done in a very short period. Experience has shown that, even when working in pairs, the students must plan their experiment very carefully, especially with respect to timing, in order to ensure a complete and accurate set of results. Since planning is essential, an outline of the practical work that will facilitate such planning now follows.

Initially, six cultures of E. coli will be started using an inoculum derived from an overnight culture grown on glycerol. These cultures will be incubated for 60–90 min. During this time the cells derived from the overnight culture, that would have been in the stationary phase of growth, will begin to enter logarithmic growth on the fresh glycerol-containing medium. After this preincubation period, during which it is essential to prepare for the rest of the experiment, two flasks will receive lactose, two IPTG, and two distilled water. It will first be necessary to remove a sample from each culture at zero time for assay of β-galactosidase and a second sample for assay of cell numbers. At 1 min the substrates will be added, and at 3, 6, 11, 21, 31, 61, and 121 min, further samples will be removed for β-galactosidase assay. In addition, at 16, 46, 76, and 121 min, samples will be withdrawn for determination of cell numbers.

The samples for enzyme assay will be taken into an "induction-stopping and membrane-dissociating mixture" consisting of chloramphenicol, which rapidly inhibits protein synthesis and hence prevents further enzyme induction, and toluene, which disrupts the cell membrane and releases the β-galactosidase quantitatively so it can be readily assayed. These samples are kept on ice for exactly 60 min to allow the enzyme to be released and the enzyme assay is then started by adding the substrate orthonitrophenolthiogalactoside (ONPG), which is made up in phosphate buffer and has been preincubated at 37°C. The assay mixture is incubated at 37°C for 10 min and the reaction, in which any β-galactosidase present hydrolizes the ONPG to release orthonitrophenol, is then stopped by adding sodium carbonate. This serves two purposes. First, it stops enzyme activity by making the pH sufficiently alkaline to totally inhibit β-galactosidase and, second, it develops the yellow coloration that is characteristic of orthonitrophenol in alkaline solution. This absorbance is determined immediately in a spectrophotometer.

The samples taken to assay growth involve adding aliquots of the culture to chloramphenicol and keeping the samples on ice. In this case, the time that the sample is kept is not crucial because the combined effect of chloramphenicol and low temperature will inhibit growth completely. Growth in all of the samples can, therefore, be determined at the end of the experiment and may be done either microscopically using a hemocytometer (or alternative form of counting chamber) or, preferably, by using an electronic cell counter (if one is available). By plotting graphs of cell number against time for each culture, it will be possible to determine the number of cells in each sample taken for β-galactosidase assay, and hence determine enzyme activity per unit cell number. Enzyme activity will be expressed as μmoles ONP released 10 min^{-1}, and this will require the preparation of a calibration curve for orthonitrophenol.

2. Materials Provided

2.1. Equipment

Stop clock
Ice bath
Vortex mixer
Waterbath at 37°C
Shaker incubator (6 spaces per pair) at 37°C
Sterile 1.0 cm^3 pipets
Bunsen burner
Test tubes
Cylinders containing bleach, or other sterilant, for used pipets
Disposal container for spent cultures
Spectrophotometer and glass cuvets
Light microscope and hemocytometer or other cell counting chamber, or an electronic cell counter.

2.2. Reagents

Overnight culture of E. coli
Flasks of minimal medium supplemented with 0.5% glycerol
0.1% Chloramphenicol
Toluene
0.1% Orthonitrophenol-β-galactoside (ONPG) in 0.05M phosphate buffer, pH 6.8

0.01M Lactose
0.01M Isopropylthiogalactoside (IPTG)
1.0M Sodium carbonate
0.0001M Orthonitrophenol (ONP)
0.05M Phosphate buffer, pH 6.8

3. Protocol

1. Having read through this entire section and the
    Introduction, in which an outline of the
    experiment is given, plan the experiment carefully
    with your partner(s). Pay particular attention to
    clearly labeling tubes and phasing the addition of
    substrate to each culture so that you do not end
    up taking more than one sample at exactly the
    same time.

    Note: *You are working with a mildly pathogenic organ-
    ism and therefore at all times respect the conventions of
    good microbiology laboratory procedure, including placing
    all used pipets in a container containing sterilant (see
    Appendix I, "Safe Handling of Microorganisms).*

2. Using aseptic technique, subculture 1.0 cm^3 of the
    overnight culture of E. coli provided into each of
    six flasks, each containing 50 cm^3 of fresh
    minimal medium supplemented with 0.5% gly-
    cerol. Clearly label each culture and incubate for
    between 60 and 90 min in a shaker, at 37°C, using
    a moderate shaking speed to ensure adequate
    aeration.

    (Note: In subsequent operations, aseptic tech-
    nique is *not* necessary because contamination of
    the cultures is unlikely to cause any problem dur-
    ing the time course of the experiment. Used
    pipets, however, should still be placed in a steri-
    lant.)

3. During the incubation period, carry out the fol-
    lowing steps.

    (a) Prepare six sets, each of eight tubes, of
        the induction-stopping and membrane-
        dissociating mixture, which contains

0.05 cm^3 chloramphenicol and 0.02 cm^3 toluene. Keep these on ice.

(b) Prepare six sets, each of eight tubes, of the "enzyme assay mixture," which consists of 2.5 cm^3 of ONPG in phosphate buffer. Keep these at $37°C$ in a water bath.

(c) Prepare six sets, each of five tubes, containing 0.20 cm^3 chloramphenicol, which will be used for the determination of cell growth. Keep these on ice.

   Note: Each set of tubes will correspond to one of the six cultures and must be clearly labeled and preferably organized in separate rows in the ice bath.

4. Remember the *planned* time schedule and stagger zero time for each culture. At zero time take 0.5 cm^3 from each culture and add it to the induction-stopping and membrane-disrupting mixture, shake on a vortex mixer, and return to the ice bath for 60 min. Also remove a 2.0-cm^3 sample into 0.20 cm^3 chloramphenicol and keep on ice for the determination of cell numbers.

5. At 1 min, add 5 cm^3 lactose to cultures 1 and 2, 5 cm^3 of IPTG to cultures 3 and 4, and 5 cm^3 of sterile distilled water to cultures 5 and 6.

6. Remove 0.5 cm^3 from each culture at 3, 6, 11, 21, 61, and 121 min into the appropriate tube of induction-stopping and membrane-dissociating mixture (*see* step 3a). Shake on a vortex mixer and leave on ice for exactly 60 min.

7. Remove 2.0 cm^3 from each culture at 16, 46, 76, and 121 min for assay of cell growth (*see* step 3c). Accumulate all these samples in the ice bath until the end of the experiment.

8. When the induction-stopping and membrane-dissociating mixture has been incubated for 60 min, add it to a tube containing 2.5 cm^3 of ONPG in phosphate buffer (*see* step 3b). Thoroughly mix and incubate at $37°C$ for 10 min.

9. To terminate the reaction and develop the yellow color of any orthonitrophenol released, add 1.0 cm^3 of sodium carbonate. Shake and determine the absorbance at 420 nm against the zero time sample for each culture, which will act as the reference blank.
Note: The amount of cellular material in each sample is so small it does not interfere with the assay. However, if for any reason this does become a problem, the sample can be centrifuged to remove debris prior to the spectrophotometric determination.

10. Determine the cell number in the various accumulated samples using either a hemocytometer (or other counting chamber) or an electronic cell counter (details of this determination are not given since this will vary according to the equipment available). If no method for determination of cell numbers is available then growth can be estimted spectrophotometrically by determination of absorbance at 660 nm. This will necessitate the prior preparation of a calibration curve relating absorbance to dry weight.

11. Prepare a calibration curve for orthonitrophenol (ONP) using the standard solution provided, which has been made up in phosphate buffer, by adding 2.5, 2.0, 1.5, 1.0, 0.5, and 0.0 cm^3 of standard to tubes containing 0, 0.5, 1.0, 1.5, 2.0, and 2.5 cm^3 phosphate buffer, respectively. Do this in duplicate. Add 1.0 cm^3 sodium carbonate to each tube, mix thoroughly on a vortex mixer, and determine absorbance of the ONP at 420 nm. Plot a graph of absorbance vs concentration of ONP.

12. Plot graphs of β-galactosidase activity (μmoles ONP released 10 min^{-1} 10^{-6} cells) against growth time taking the average result of the two replicate cultures for each treatment.

13. Carefully dispose of the spent cultures by placing them in the disposal container ready for autoclaving.

4.   Results and Discussion

Typical results for this experiment are shown in Fig. 1. These show that β-galactosidase is rapidly induced by the introduction of lactose and IPTG into the growth medium,

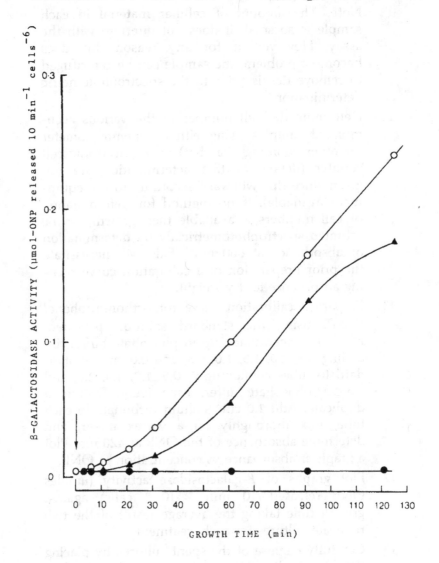

Fig. 1. The time course of induction of β-galactosidase in cultures of E. coli grown on glycerol (●), lactose (▲), and IPTG (○).

the rate of induction being greater and to a higher level in the presence of the gratuitous inducer. Glycerol-grown cultures contain no detectable β-galactosidase. It should be remembered that the results allow for growth of the cultures during the experiment by expressing β-galactosidase activity, 10^{-6} cells. It is also useful to plot graphs of growth of the cultures since these will show that the IPTG and glycerol cultures grow slowly, whereas the rate of growth in the lactose-containing culture increases as the β-galactosidase is induced and the cells begin to utilize lactose as the preferred growth substrate to glycerol. This indicates that IPTG, although causing induction, is not being used as a growth substrate.

The reason for the more rapid rate and higher level of induction caused by IPTG compared to lactose can only be surmised from the experimental data. There are various possible explanations, including: (a) more rapid uptake of IPTG than lactose; (b) greater affinity of IPTG than that of lactose (or its metabolite allolactose, which is believed to be the natural inducer) for the lac repressor protein that controls gene expression (1–4) (c) the fact that IPTG is not metabolized may lead to a lower level of ATP (higher cAMP) in these cells compared to cells metabolizing lactose, and cAMP is known to have an effect on the transcription of the lac mRNA. These, and other possibilities, could be examined further in additional experiments that might include:

(a) Determination of lactose and IPTG uptake from the growth medium using spectrophotometric assays for the determination;
(b) Determination of cAMP levels in cells growing in the three different conditions;
(c) Examination of the effects of transcriptional and translational inhibitors on the induction process (suitable inhibitors would be rifampicin, chloramphenicol, or tetracycline).

References

1. Alberts, B., Bray, D., Lewis, J., Raff, M., Roberts, K., and Watson, J.D. (1983) *Molecular Biology of The Cell.* Garland, New York, pp. 435–455.

2. Lehninger, A.L. (1982) *Principles of Biochemistry*. Worth, New York, pp. 871–912.
3. Stryer, L. (1981) *Biochemistry* (2nd Ed.) Freeman, San Fransisco, pp. 669–684.
4. Zubay, G. (1982). *Biochemistry*. Addison Wesley, Reading, Massachusetts, pp. 979–1016.

Chapter 23

Experiments on Enzyme Induction in the Unicellular Green Alga *Chlorella fusca*

Kenneth H. Goulding

1. Introduction

The ability of microorganisms to adapt their enzyme complement in response to varying growth conditions has been established since the early years of this century. The experiments reported here use the unicellular green alga, *Chlorella fusca*, in a study of the induction of the enzyme isocitrate lyase in response to the presence of acetate in the growth medium in a variety of growth conditions. These include variation in the light and aeration regimes, and the presence and absence of glucose. The experiments are essentially simple and do not employ sophisticated techniques. However, they do represent a different approach to conventional experiments on enzyme induction that, almost invariably, use the bacterium *Escherichia coli* as the test organism in a straightforward examination of the time course of induction of β-galactosidase in response to lactose. In addition to the "basic" experiment, a wide range of additional experiments are suggested based on the same general approach. This enables the generation of experimental data that are either relatively straightforward to interpret, or data that can tax an advanced student. The experiments outlined are suitable for a practical class of 3 h or can be extended to provide a project that might occupy a student, or group of students, for 10 or more 6-h practical sessions.

Many microorganisms can grow upon acetate as the sole carbon substrate. This usually involves induction of two enzymes, isocitrate lyase (EC 4.1.3.1) and malate synthetase (EC 4.1.3.2). These enzymes are crucial to the ability to grow on acetate and constitute the so-called glyoxylate cycle (glyoxylate bypass) of the tricarboxylic acid (TCA, or Krebs) cycle.

As shown in Fig. 1, acetate, which enters the TCA cycle as acetyl CoA by condensation with oxaloacetate to form citrate, is, as a result of one turn of the cycle, completely oxidized to carbon dioxide, whereas the oxaloacetate acceptor is regenerated. At various stages in this oxidative process, electron acceptors are reduced. These are reoxidized via electron transport systems, and ATP is generated. The chemical energy so conserved provides for cellular maintenance and growth. However, if the organism is to grow upon acetate, the TCA cycle must not only function as a *catabolic*, energy-yielding pathway, but must also provide biosynthetic precursors for the synthesis of amino acids (and hence proteins), fatty acids (and hence lipids), carbohydrates, and other cellular components (Fig. 1). This dual function, or as it is more properly known, *amphibolic* role, of the TCA cycle can only be satisfied if carbon is continually fed into the cycle to maintain the level of its intermediates. Thus, there is a requirement for an *anaplerotic* pathway that feeds carbon from the substrate acetate, into the cycle. As outlined above, the TCA cycle itself completely oxidizes acetate to carbon dioxide, with no net increase in levels of intermediates and, hence, no provision for the supply of biosynthetic precursors. The glyoxylate cycle, however, circumvents the two oxidative decarboxylations of the TCA cycle (Fig. 1). Isocitrate lyase catalyses the split of the six-carbon isocitrate to the four-carbon suc-

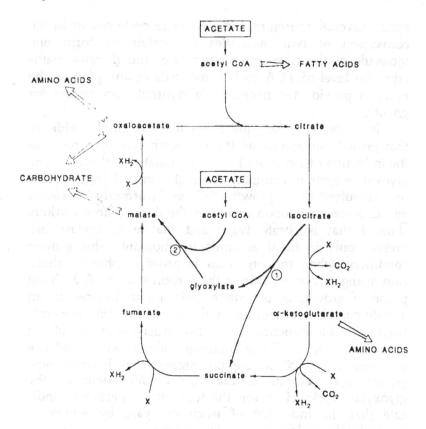

Fig. 1. The tricarboxylic acid (TCA) cycle and glyoxylate bypass. The reactions of the TCA cycle, which lead to the complete oxidation of acetate to two molecules of carbon dioxide and the reduction of electron acceptors (X → XHsub2) are shown by ⇒. The anaplerotic reactions of the glyoxylate cycle are catalyzed by isocitrate lyase (1) and malate synthetase (2). The sites of withdrawal of intermediates as precursors for synthesis of major cellular compenents are shown by ⇒. For full explanation, *see* text.

cinate and the two-carbon keto acid, glyoxylate. This condenses with a second molecule of acetyl CoA in a reaction catalyzed by malate synthetase that results in the production, from the two two-carbon units, of the four-carbon compound, malate. This is used to regenerate the oxaloacetate acceptor molecule via the normal reaction of the TCA cycle. Overall, therefore, the glyoxylate cycle results in the conversion of two molecules of acetate to form one molecule of the four-carbon succinate, and thereby maintains the level of TCA cycle intermediates and permits the cycle to provide the necessary biosynthetic precursors for growth.

In bacteria, yeasts, and fungi there is ample evidence that growth on acetate as the sole carbon source requires the induction of isocitrate lyase and malate synthetase. The glyoxylate cycle has similarly been shown by Syrett et al.(1) to be involved in the growth of C. fusca (formerly C. vulgaris and C. pyrenoidosa) upon acetate in the dark. These workers showed that isocitrate lyase and malate synthetase are present only in basal amounts in photoautrophic growth conditions when the organism is growing photosynthetically using carbon dioxide as the carbon source. A 24-h lag phase of growth occurs when cells are transferred to an acetate-containing medium in the dark, and this is correlated with the induction of the two crucial enzymes of the glyoxylate cycle. These findings, along with evidence obtained using ^{14}C-acetate (2) suggest that heterotrophic growth upon acetate in the alga is dependent on the glyoxylate cycle. However, the following experiments indicate that the induction of isocitrate lyase by acetate is markedly dependent on growth conditions.

In the following sections, details are given for an experiment in which the induction of isocitrate lyase in cultures of C. fusca, grown in a variety of experimental conditions, is examined. This may be regarded as the "basic" experiment. Many further experiments are possible using the same approach, some of which are suggested in section 4.

2. Materials Provided

2.1. Equipment

Ice bath
Stopclock
Waterbath at 30°C
Spectrophotometer and glass cuvets
Test tubes

Centrifuge tubes
Oven at 105°C
Desiccator

2.2. Reagents

0.067M Phosphate buffer, pH 7.5
0.15M Magnesium Sulfate
0.24M Trisodium isocitrate
0.06M Reduced glutathione
15% Trichloroacetic acid
0.1% Dinitrophenylhydrazine in 2M hydrochloric acid
2.5M Sodium hydroxide
0.004M Sodium glyoxylate standard

2.3. Enzyme Preparations

Use enzyme preparations from cells grown in cultures A–H, under the following conditions:
A. Light + CO$_2$
B. Light − CO$_2$
C. Light + CO$_2$ + acetate
D. Light − CO$_2$ + acetate
E. Dark + CO$_2$
F. Dark + CO$_2$ + acetate
G. Dark + CO$_2$ + glucose
H. Dark + CO$_2$ + acetate + glucose

Details of cell growth and the method for producing enzyme preparations are given in Appendix II, Chapter 23.

3. Protocol

The assay of isocitrate lyase in each of the eight cultures (A–H) provided depends on the conversion of glyoxylate, formed by the action of the enzyme on isocitrate, to a colored dinitrophenylhydrazone derivative by reaction with dinitrophenylhydrazine. Results are expressed as μmol glyoxylate formed h^{-1} mg dry weight cells^{-1}. Thus, it is necessary to prepare a calibration curve for glyoxylate and also to determine the dry weight of each culture used.

1. Slowly thaw out the cultures provided and keep on ice until required.

2. Take duplicate 1.0 mL samples of each culture, which should be shaken to ensure homogeneity, into preweighed, labeled centrifuge tubes.

3. Centrifuge at 500g for 5 min. Carefully remove the supernatant and place the tubes containing the pellets in an oven at 105°C. Leave overnight. Place the tubes in a desiccator to cool. Reweigh.

4. Using the reagents provided, add the following to each of 16 test tubes numbered A1, A2–H1, H2.

    1.0 mL Phosphate buffer, pH 7.5
    0.1 mL Magnesium sulfate
    0.1 mL Sodium isocitrate
    0.1 mL Reduced glutathione
    1.7 mL Distilled water

5. Set up eight more tubes identical to those in step 4, with isocitrate omitted and the amount of distilled water increased to 1.8 mL. Label these tubes A3–H3. (These act as blanks and allow for any keto acid in each sample that will produce a phenylhydrazone and, therefore, could give false positive results.)

6. Place the tubes in a waterbath at 30°C to equilibrate for at least 5 min.

7. Start a stopclock and at zero time add 0.5 mL of culture A to tube A1. Shake and return to the waterbath at 30°C. After 1 min, add another 0.5 mL of culture A to A2 and, after another minute, add 0.5 mL to the blank A3. Repeat this procedure with cultures B–H until the reaction has been started in all 24 tubes.

8. Allow each reaction to proceed for 60 min, occasionally shaking the tubes to prevent settling out of the cells.

9. During the incubation period, prepare, at room temperature, 24 tubes labeled A1, A2, A3–H1, H2, H3, each containing:

    0.4 mL Trichloroacetic acid
    0.8 mL Dinitrophenylhydrazine

    (In this stopping/developing mixture, the trichloracetic acid serves to stop the enzyme reaction and precipitate out the protein in solution, whereas the dinitrophenylhydrazine reacts with any glyoxylate present to form glyoxylate dinitrophenylhydrazone.)

10. At the end of the 60-min reaction period, carefully shake the tube and remove a 1.0-mL sample into the correspondingly labeled stopping/developing mixture. Shake and incubate at 30°C for 15 min, during which time the reaction to form the phenylhydrazone will go to completion.

11. To develop the color of the phenylhydrazone, add 3.2 mL sodium hydroxide. The color is stable, so it is best to wait until all the samples are ready before proceeding to the next stage.

12. Remove precipitated protein and cell debris by centrifugation at 500g for 5 min.

13. Read the absorbance at 445 nm using the appropriate blank (A3–H3) for each culture to zero the spectrophotometer.

14. Prepare a calibration curve for glyoxylate between 0 and 4.0 μmol mL^{-1} by adding 0, 0.2, 0.4, 0.6, 0.8, and 1.0 mL of the standard glyoxylate solution to each of six more stopping/developing tubes, as in step 9. Adjust the volume in each tube by adding 1.0, 0.8, 0.6, 0.4, 0.2, and 0.0 mL of distilled water, respectively. Incubate at 30°C for 15 min, add 3.2 mL sodium hydroxide, mix, and determine the absorbance at 445 mn. Plot a calibration curve of absorbance vs concentration of glyoxylate.

15. Using the calibration curve and the dry wt for each culture as determined in steps 1–3, express the results as μmol glyoxylate formed h^{-1} mg dry weight cells^{-1} for each of the cultures.

4. Results and Discussion

Typical results for the "basic" experiment described in section 3 are given in Table 1. The only conditions in which isocitrate lyase is induced are those in which acetate acts as the sole carbon source, as in cultures D and F. Cul-

Table 1
The Effect of Growth Conditions on the Induction
of Isocitrate Lyase in *Chlorella fusca*

Culture	Growth conditions	Isocitrate lyase activity,[a] μmol glyoxylate h^{-1} mg dry wt
A	Light + CO_2	0.03 ± 0.01
B	Light − CO_2	0.05 ± 0.02
C	Light + CO_2 + acetate	0.04 ± 0.01
D	Light − CO_2 + acetate	0.64 ± 0.04
E	Dark + CO_2	0.03 ± 0.01
F	Dark + CO_2 + acetate	0.68 ± 0.05
G	Dark + CO_2 + glucose	0.03 ± 0.01
H	Dark + CO_2 + acetate + glucose	0.06 ± 0.02

[a] Figures show means from six separate experiments.

tures A (light + CO_2) can grow autotrophically and no acetate is present to act as inducer, whereas culture B (light − CO_2) is incapable of any growth since no carbon substrate is available. Cells grown as in C (light + CO_2 + acetate) have two "choices" of carbon substrate. They may either grow photosynthetically using the CO_2 or they may utilize the acetate by inducing the glyoxylate cycle. As shown in Table 1, isocitrate lyase is not induced in these conditions, thus indicating a preference for autotrophic growth. In the absence of CO_2, however, cells grown on acetate in the light (D) do induce isocitrate lyase, indicating that, when the algae are forced to use acetate as the sole substrate, they induce the glyoxylate cycle in order to provide the necessary biosynthetic precursors for growth. An alternative strategy would be to completely oxidize the acetate via the TCA cycle to carbon dioxide that could then be refixed photosynthetically. This oxidation followed by reduction is clearly more inefficient than the glyoxylate cycle, and this is presumably why the glyoxylate cycle is induced.

In the dark, no growth can occur in the absence of an organic substrate. Culture E (dark + CO_2) therefore acts as a sort of control indicating the basal level of the enzyme. When forced to use acetate as the sole growth substrate as in culture F (dark + CO_2 + acetate), isocitrate lyase is induced. However, no induction occurs in culture G (dark + CO_2 + glucose), when glucose is the sole substrate, since in these conditions, isocitrate lyase is not necessary to provide the biosynthetic precursors required for growth. Clearly, glucose, when available, acts as the preferred substrate in the dark since, as in culture H (dark + CO_2 + acetate + glucose), it represses the induction of isocitrate lyase even when acetate is present.

Thus the "basic" experiment indicates that growth conditions are crucial to the induction of isocitrate lyase in *C. fusca*. It is not simply induced by acetate; indeed, from the results given, it would appear that *only* when acetate is the *sole* growth substrate is the enzyme induced.

As indicated in the Introduction, the "basic" experiment is an unusual demonstration of enzyme induction in a unicellular microorganism that provokes more thought, in terms of data interpretation, than many other experiments on enzyme induction. Moreover, the "basic" experiment can be adapted in many ways to provide material for a wide variety of further experiments. Below are some suggestions.

1. Examine the time course of induction and rate of growth when cells are growing on acetate as sole carbon substrate in the light or dark (growth conditions D and F). In this case, grow cells as in Appendix II, Chapter 23, but take samples 6, 12, 24, 36, and 48 h after addition of acetate for enzyme assay and determination of growth. (Growth may be determined either by cell counting or by determining absorbance at 665 nm.)

2. Examine the effect of metabolic inhibitors on the induction process by carrying out one or more of the following experiments.

   (a) Add the herbicide Diuron (dichlorophenyldimenthylurea, or DCMU) ($10^{-6}M$), which inhibits noncyclic photophosphorylation, to cultures grown in condition C (light + CO_2 + acetate) that do not normally support the induction of isocitrate lyase. It will be shown that inhibiting photosynthetic growth causes the cells to utilize acetate by inducing isocitrate lyase.

(b) Add the inhibitor of protein synthesis, chloramphenicol ($10^{-4}M$), to cultures grown in condition D (light $-$ CO_2 $+$ acetate) or F (dark $+$ CO_2 $+$ acetate) that normally do result in induction. No induction occurs, indicating that *de novo* protein synthesis is involved in the induction process.

(c) Add the uncoupler of oxidative phosphorylation, dinitrophenol ($10^{-4}M$), to cells grown in condition F (dark $+$ CO_2 $+$ acetate). This will result in induction being inhibited, thereby showing energy is required for the induction process. If the same experiment is repeated in condition D (light $-$ CO_2 $+$ acetate), dinitrophenol will have no effect, since it does not inhibit photophosphorylation.

3. Use radioactively labeled ^{14}C-acetate to follow the distribution of acetate into cellular fractions (released CO_2, water soluble substances, fatty acids and lipids, polysaccharides and proteins) in induced and non-induced cells in both the light and dark (for details of such experiments, *see* ref. 3). These experiments, which can be simple or quite extensive, will show that much more acetate is incorporated into cellular material in induced cells than in non-induced cells.

References

1. Syrett, P.J., Merrett, M.J., and Bocks, S.M. (1963) Enzymes of the glyoxylate cycle in *Chlorella vulgaris*. *J. Exp. Bot.* **14**, 249–64.
2. Goulding, K.H. and Merrett, M.J. (1966) The photometabolism of acetate by *Chlorella pyrenoidasa*. *J. Exp. Bot.* **17**, 678–689.
3. Goulding, K.H. and Merrett, M.J. (1967) Short-term products of ^{14}C-acetate assimilation by *Chlorella pyrenoidosa* in the light. *J. Exp. Bot.* **18**, 128–139.

Chapter 24

Yeast Protoplast Fusion

Elliot B. Gingold

1. Introduction

A major limitation to the results that could be achieved using traditional methods in the study of microbial genetics has been the barriers to fusion that exist both between isolates belonging to different species, and also often between different isolates of the same species. Thus a scientific worker may have obtained two varieties of fungi, each with their own desirable characteristics, but have been unable to combine the characteristics in a single organism. Even more severe barriers to genetic recombination were often found when working with bacteria.

During the past ten years, a novel approach has been used to overcome such fusion barriers. If the cell wall is removed, one is left with what is known as a protoplast. Such cells are fragile and require careful handling, but most importantly they are viable and capable of being regenerated into the normal-walled organism. While in the protoplast state, however, genetic manipulations not possible with complete cells, can be carried out. In particular, protoplasts can be fused with the help of agents such as polyethylene glycol (PEG), in conjunction with Ca^{2+} ions. It is thus possible to remove the cell walls from two organ-

isms normally incapable of fusing, induce fusions in the
protoplast state, and regenerate the new combined organism by allowing cell walls to reform. Such protoplast
fusions can facilitate crosses between both incompatible
members of the same species, and also between members
of different species. A good review of the scope of this
technique in both bacteria and fungi can be found in
Ferenczy (1).

In this experiment, the technique of protoplast fusion
is used to bring together two normally incompatible strains
of the yeast *Saccharomyces cerivisiae*. This yeast has two mat-
ing types, a and α, and under normal conditions fusions are
only possible between strains of opposite type. By working
with protoplasts, such barriers are removed, and in this
exercise high levels of fusion between strains of the same
mating-type should be achieved.

The first stage in the procedure is the removal of the
yeast cell walls and the consequent release of protoplasts.
It has been found effective to first weaken the walls by use
of the strong reducing agent, dithiothreitol. Following this
treatment, an enzyme preparation can be used to remove
the wall. In this procedure we use the microbial enzyme
Novozym 234; other enzymes such as that found in snail
gut juice have also proved effective.

Following careful washing of the protoplasts, fusion is
achieved by the standard technique of mixing the strains in
a PEG–Ca^{2+} solution. To regenerate yeast cell walls, protoplasts must be embedded in solid medium. Thus the
fusion mix is added into warm, soft agar medium that is
then poured into petri dishes and allowed to set. Each of
the parental strains has different nutritional requirements,
and hence use of a medium on which only a recombinant
cell can grow allows selection of the fusion products.

It is of interest to realize that fusion is not the only
manipulation possible with protoplasts. Using very similar
procedures it is found that protoplasts will take up DNA
from the PEG–Ca^{2+} solution. This technique is the basis
of the rapidly expanding field of yeast genetic engineering.
If a plasmid capable of being propagated in yeast is available, the present exercise can be modified to demonstrate
such yeast transformation (*see* Appendix II, Chapter 24).

Overall, the exercise requires no special equipment and takes around 5 h. Time is available within this period for discussion or other work. In addition, it will be necessary for students to return to the laboratory 4–5 d later to view results.

2. Materials Provided

2.1. Strains

Overnight culture of BSC465/2a (genotype a, *ade2 his3 ura3*)
Overnight culture of FM811c (genotype a, *ade2 lysl*)

2.2. Solutions

Sterile distilled water
Reducing buffer
1.2M Sorbitol
Enzyme buffer
Enzyme preparation (Novozym 234)
1% SDS
Sorbitol–CaCl$_2$ solution
PEG–CaCl$_2$ solution

2.3. Media

Molten minimal agar
Adenine, histidine, uracil, and lysine supplements

2.4. Equipment

Two sterile, disposable, plastic centrifuge tubes
Sterile 10- and 1-mL pipets
Bench centrifuge (refrigerated or in cold room)
Ice bucket
30°C Waterbath
45°C Waterbath
Microscope
Small glass tubes
Sterile glass universals
30°C Incubator

3. Protocol

You are provided with overnight cultures (around 10^8 cells/mL) of two yeast strains in YEPD. In this protocol it is assumed that Parent A is BSC465/2a and Parent B is FM811c. If other strains are being used, it will be necessary to alter the nutrient supplements for the regeneration step.

Remember throughout this procedure that protoplasts are fragile. For this reason they are kept in osmotically stabilized solutions (1.2M sorbitol). It is essential to handle them with care; centrifuge speeds are kept low and resuspensions are gently performed. If possible, centrifugation steps are performed in a refrigerated bench centrifuge or in a cold room; take care to keep all solutions ice cold.

3.1. Protoplast Formation

1. Harvest 10 mL aliquots of each culture by spinning in plastic, disposable centrifuge tubes in a bench centrifuge. A short (1-min) spin should sediment the cells without difficulty.

2. Remove the supernatant and resuspend the cells in 10 mL of cold sterile water. The cells are then recentrifuged and the supernatants discarded.

3. Resuspend the cells in 4 mL of reducing buffer. Place the tubes in a 30°C waterbath for 10 min, occasionally shaking the tubes to keep the cells suspended.

4. Centrifuge as before, resuspend the pellets in 10 mL 1.2M sorbitol, and recentrifuge. Repeat this washing step.

5. From this point in the procedure protoplasts will be formed and the warning about careful handling must be remembered. Resuspend the pellets in 3 mL of enzyme buffer and add 1 mL of the enzyme preparation. The suspensions are incubated at 30°C with gentle shaking.

6. It is desirable to follow the progress of the protoplasting microscopically. This is because some strains will form protoplasts more rapidly than

others. In addition, leaving strains too long in the protoplasting solution often reduces their later regeneration. Observe the protoplasts with a magnification of $\times 400$. Light field microscopy will be adequate, although use of phase contrast provides a clearer view of the actual protoplasts.

Prepare two series of tubes, one with 0.5 mL 1.2M sorbitol in each tube and the other with 0.5 mL 1% SDS in each tube. At frequent intervals (10–15 min) withdraw 0.05-mL aliquots from the protoplasting suspensions and mix into the sorbitol and SDS. Observe the results microscopically. Throughout the process the number of cells observed in the sorbitol should not change, although they will become rounder and less distinctly outlined. As protoplasting proceeds, however, the numbers observable in the SDS should drop, since protoplasts will lyse in this solution.

When the cultures reach the point at which cells no longer appear in the SDS samples, remove the protoplast suspensions to ice. In general, up to 60 min incubation will be required to achieve full protoplasting.

7. Gently centrifuge the protoplast suspensions (1000g for 10 min should be adequate for this and all subsequent spins). Carefully resuspend the pellets in 10 mL of cold 1.2M sorbitol. Recentrifuge to pellet protoplasts.

8. Repeat the washing procedure in step 7. Resuspend the pellets in 1.0 mL sorbitol–$CaCl_2$.

3.2. Protoplast Fusion

1. Mix the two suspensions, centrifuge, and resuspend the pellet in 1.0 mL of the PEG–$CaCl_2$ solution. Leave at room temperature for 20 min. It is worthwhile observing the mixture microscopically at this stage. The dehydrating and aggregating nature of the PEG treatment should be very apparent.

2. Pellet the protoplasts and resuspend in 2 mL sorbitol–$CaCl_2$ buffer.

3.3. *Protoplast Regeneration*

1. Prepare the following molten agar bottles:
    (a) 50 mL minimal agar, 0.05 mL adenine supplement
    (b) 50 mL minimal agar, 0.05 mL adenine supplement, 0.05 mL histidine supplement, 0.05 mL uracil supplement
    (c) 50 mL minimal agar, 0.05 mL adenine supplement, 0.05 mL lysine supplement.

    Medium a will allow fusion diploids to grow, medium b will allow Parent A to grow, and medium c will allow Parent B to grow.

2. Distribute 4×10 mL of medium a to four prewarmed universals, repeat with 4×10 mL of media b and c. Store in a 45°C waterbath.

3. Prepare a series of tenfold dilutions of the fusion mixture in sorbitol–$CaCl_2$ solution. The dilutions should be 10^{-1}, 10^{-2}, 10^{-3}, 10^{-4}, and 10^{-5}.

4. Aliquots of 0.1 mL of the dilutions are mixed with the molten media and poured into empty sterile petri dishes. For medium a, plate an aliquot of the undiluted, 10^{-1}, 10^{-2}, and 10^{-3} diluted samples; for media b and c, plate the 10^{-2}, 10^{-3}, 10^{-4}, and 10^{-5} samples.

5. Incubate the plates at 30°C. Observe after 4–5 d and count colonies on the plates. If there are too many to count on a given plate, record this information.

4. Results and Discussion

From the counts on medium a, calculate the total number of fusion products in the undiluted fusion mix. Repeat this step for the results from media b and c, calculating the total number of Parent A and Parent B cells that were able to regenerate.

If the experiment has worked well, one can get up to 10^6 fusion products/mL and 10^8 regenerent parental cells/

mL for each parental type. This can be compared to the theoretical figure of 5×10^8 cells/mL of each type in the 2 mL of fusion mix.

In general, results of this magnitude are not obtained, especially at the first attempt. It is for this reason that a number of dilutions are plated. The explanations for the apparent reduction are many, but include:

(a) Loss of cells during centrifugation
(b) Damage of protoplasts during handling
(c) Protoplasts remaining clumped after PEG treatment, thus further reducing the apparent plate count.

Nonetheless, it is usual to get a reasonable proportion of the measured regenerent colonies as fusion products. Results better than the 1% fusion described above are perfectly feasible.

References

1. Ferenczy, L. (1981) Microbial Protoplast Fusion, in *Society for General Microbiology*, Symposium 31, "Genetics as a Tool in Microbiology," (Glover, S.W. and Hopwood, D.A., eds.) Cambridge University Press, Cambridge, UK.

Chapter 25

The Ames Test

Elliot B. Gingold

1. Introduction

A mutagen is any treatment or chemical substance that can induce heritable changes in genetic material and thereby increase mutation rates. For many decades scientists have been making use of such mutagens with aims as diverse as, on the one hand, increasing yields in strains used for commercial antibiotic production, to, on the other hand, the search for the variants that assist molecular biologists in understanding fundamental cellular processes. For some time, however, it has been apparent that as well as being useful tools, mutagens can be dangerous human health hazards. In particular, a high proportion of mutagens have been shown to have carcinogenic activity. Conversely, 85% of known human carcinogens were found to be mutagenic when tested on bacteria.

The clear link between carcinogenic and mutagenic activity has been put to good use in the Ames Test. Direct carcinogenicity testing is a long and expensive procedure, since it involves extensive use of laboratory animals. It would be totally unfeasible to test the vast range of compounds we are exposed to using such animal tests. Mutagenicity tests, however, can be performed on microbial systems and hence are quick and cheap. It is thus possible to screen large numbers of compounds for mutagenicity and obtain generally reliable indications of potential carcinogenicity. In this way, human health hazards from substances ranging from biocides to hairdyes have been revealed. In many cases, workers have subsequently confirmed the carcinogenic activities using animal tests. It must be emphasized, however, that without the preliminary microbial mutagenicity screen, few of these substances would have ever been suspected.

By far the most used microbial test system is that developed by Bruce Ames and his colleagues (1). As a result of other studies, this laboratory had obtained a range of histidine biosynthesis mutants from *Salmonella typhimurium* and were able to use these to develop a sensitive test for mutagens. The basis of the test is the measurement of the reversion rate $his^- \rightarrow his^+$; that is, of histidine-requiring strains back to the wild-type ability to synthesis histidine. Spontaneous reversion to his^+ is not a high-frequency event, mutation such as the forward $his^+ \rightarrow his^-$ occurs with far higher rates, but it has the advantage of being easily and accurately measured. To detect revertants, it is necessary only to plate samples of the culture onto a minimal medium without added histidine. On such a medium only revertants can grow into colonies. Since a 0.1-mL sample of culture may contain up to 10^8 cells, even low numbers of revertants may be detected and the reversion rate measured. If treatment with a test compound increases the reversion rate, this can be easily seen. Thus, this provides a sensitive test for mutagenic action.

To make reversion possible, all his^- mutations used in the Ames test are point mutations; that is, they differ from the wild-type by a single base change. A reversion is thus the exact opposite change to the original mutation. But since different mutagens have different actions, more than one test strain is needed. Some mutagens act by *substituting* one base pair for another, thereby inducing a *base substitution* mutation. Other mutagens act on the DNA helix to either *add* or *delete* a single base pair; such mutagens are said to produce a *frameshift* mutation. Now consider a test using only a single strain, a his^- mutant resulting from a base substitution. When treated with a mutagen that acts by base substitution, mutations of this type will occur in different cells in a wide range of genes. Among these many new mutants, however, will be some cells in which the new substitution has simply restored the original his^- change back to its wild-type form. The formation of these induced revertants would cause an increase in the measured reversion rate, and thus the mutagenic activity would be detected. But consider the effect of treating this strain with another mutagen, one that has a frameshift action. There is no way that its action of adding or deleting a base pair can correct an error caused by an incorrect base pair. Hence no revertants are induced and the compound would appear to give a negative result. To have detected this mutagen, a second test strain would have been required, again his^-, but this time as a result of a frameshift mutation.

From this discussion it would appear that two test strains would be required, one his^- caused by a base substitution and the second his^- caused by a frameshift. In actual practice, however, it has been found necessary to include two different strains with frameshift mutations. This complication arises from the fact that different frameshift mutagens appear to be specific for one of two types of DNA base sequences (1). Thus Ames developed his test around three his^- mutations, a base substitution, found in strain TA1535, and two different frameshifts, found in strains TA1537 and TA1538.

The great sensitivity of the Ames test is the result of a number of additional features of the test strains. In the first place, all the strains have the *rfa* allele that results in partial loss of the lipopolysaccharide barrier that coats the surface of wild-type cells. As a result of this loss, many substances that would otherwise have been blocked can enter the cell. Obviously, this is necessary to test for mutagenic activity. In addition to this gene, the test strains are also unusual in having a deletion that removes the gene *uvrB*, and hence are unable to use the major error-free pathway for repair of DNA damage, excision repair. Many mutagens actually act indirectly by causing DNA damage that must then be repaired if the cell is to remain viable. In such cases, the mutation that is observed is caused by misrepair. In the test strains, the only pathways available for repair are error-prone, and hence the rate of misrepair is high. For this reason the test strains are particularly sensitive to these indirect-acting mutagens.

Still greater sensitivity to many mutagens has been obtained by the addition of an R plasmid (carrying antibiotic resistance) to the standard strains. This increased sensitivity has, in fact, nothing to do with the antibiotic resistance genes themselves, but arises from another gene on the plasmid. The two additional strains generally used are TA100 (TA1535, plus the plasmid) and TA98 (TA1538, plus the plasmid). The standard Ames test, in fact, consists of the five strains TA100, TA98, TA1535, and TA1538. More recently, the strain TA97 has been introduced as a more sensitive replacement for TA1537 (2).

Even the most sensitive bacterial strain possible would completely fail to detect the large group of compounds that are themselves nonmutagenic, but are converted by mammalian enzyme systems into mutagenic and hence potentially carcinogenic metabolities. Because bacteria do not possess similar enzymes, these compounds would be presented to the bacterial genetic system in their nonmutagenic form. To avoid such "false negative" results, a simple modification has been made to the standard test—the addition to the test medium of a rat liver microsomal fraction. To obtain maximum enzyme activity, the rats are first exposed to compounds known to induce such activity. The rats are then killed, the livers removed and homogenized, and the suspension centrifuged at 9000g. The supernatant, called the "S-9 fraction," is collected and stored frozen. Before use it is mixed with a number of cofactors to give the S-9 mix, and then added to the test system. In this way the compounds to be tested are exposed to the same enzymic activities found in a mamalian system and, hence, mutagen activation can occur.

The test procedure itself is very simple; to a base layer of Vogel-Bonner (minimal) medium, a soft agar overlay is added. Before pouring, a number of additions are made to this overlay agar: the test bacteria, some biotin, a trace of histidine, and, where relevant, the S-9 mix. The actual test compound may also be added at this stage, or as in the procedure to be followed here, spotted onto the agar surface after it has set. The plates are then incubated at 37°C for 2–3 d. The biotin is added since this is a growth requirement for these strains. The trace of histidine is added to allow the *his⁻* test bacteria to grow for a small number of generations. This growth can be seen as a cloudy lawn, with only *his⁺* revertants growing into full colonies. The main reason this growth is required is that most mutagens only act on dividing cells and thus give only positive results in the presence of some growth. However, the background lawn itself provides another useful function; it demonstrates that a nontoxic dose of the test compound has been used. Similarly, the absence of a lawn indicates a toxic dose. This can be very important since many compounds show a strong activity only close to (but obviously below) the toxic dose; the appearance of the lawn can thus provide a useful indication of whether the correct range of doses is being tested. Of course, in a spot test such as the present exercise, a concentration gradient is set up around the spot and a range of concentrations is thus tested on the one plate. In such tests it is common to see a clear area immediately around the spot surrounded by the cloudy lawn. When a positive result is achieved, it will generally be seen as a ring of revertant colonies surrounding the test spot and just outside the toxic zone.

In this exercise, three of the strains, TA100, TA98, and TA1537, will be used. For each of the strains, a number of plates will be set up. The first plate will have no test compound added and will thus give a measure of the spontaneous reversion rate. A similar control with no test compound, but with S-9 in the overlay, will indicate if this addition has any effect on the number of colonies. (A large change should not be seen.) Further plates will have known mutagens spotted onto the agar. These are positive controls, set up to check if the system is operating correctly. One of these control mutagens is tested both with and without S-9 addition. This is a compound requiring activation; these plates are thus included to test whether the S-9 is functioning. Finally, plates with the test substance are prepared both with and without S-9 mix.

In this experiment it is possible to test a wide range of substances, such as laboratory chemicals, biocides, cigaret smoke condensate, and household products. It was in such a laboratory class at Berkeley that the mutagenic activity of hair dyes was first discovered. Any unexpected positive results should, of course, be followed up. This could take the form of a project using the full Ames protocol (2), fol-

lowed by publication or, alternatively, by notifying a
laboratory working in this area of your results.

2. Materials Provided

2.1. Cultures

Overnight cultures of TA100, TA98, and TA1537 are used.

2.2. Media

Vogel-Bonner (minimal) plates
Molten overlay agar (45°C in the water bath)
Universal bottles (45°C)

2.3. Additives

Histidine–biotin supplement
S-9 mix

2.4. Mutagens

Methyl methane sulfonate (MMS)
9-Aminoacridine
2-Aminoanthracene
Test substance

2.5. Equipment

45°C Waterbath
37°C Incubator
Micropipet with sterile tips (0–20 μL)
Sterile, 1 and 10 mL pipets
Pipet filler
Disposal vessels for contaminated glassware
Disposal beaker for micropipet tips

3. Protocol

1. In performing this experiment, it is important to
    be fully aware of safety. *Salmonella typhimurium* is

a well known cause of food poisoning, and
although the strains used in this practical class are
of very low virulence, it is nonetheless wise to
take every precaution with the cultures. Safety
precautions pertaining to the use of microorganisms are listed in Appendix I. Sterile technique
must be used throughout and all wastes placed in
the containers provided for autoclaving. Mouth
pipeting should be avoided throughout the exercise. Potentially more dangerous are the
mutagens; when handling these, every care should
be taken to avoid spills or other accidents.
Mutagens will be dispensed by micropipets; the
tips should be disposed of in the specially marked
containers.

2. Before commencing any practical work, plan what
   you are to do. It is at this stage that plates are
   labeled. Divide the plates into three groups of
   eight, each group corresponding to one of TA100,
   TA98, or TA1537. Within each group, the eight
   plates should be labeled with the strain and the
   following:
   - (a) Control
   - (b) Control + S-9
   - (c) Methyl methane sulfonate (MMS)
   - (d) 9-Aminoacridine
   - (e) 2-Aminoanthracene
   - (f) 2-Aminoanthracene + S-9
   - (g) Test compound
   - (h) Test compound + S-9

3. To the 50 mL of molten top agar, add 5 mL of the
   histidine–biotin supplement. Mix by gently swirling and return to the 45°C waterbath.

4. Pipet 2 mL of the molten agar into each of 24
   small bottles, taking care to maintain sterile conditions. Return each to the waterbath immediately.

5. To pour an overlay, remove one bottle from the
   waterbath, carefully wipe dry with tissue, and add
   0.1 mL of the appropriate strain. In cases where
   S-9 is required, add 0.5 mL of the S-9 mix.
   Quickly mix and pour onto the plate. Spread the

overlay by gentle rotation. It is important not to spend too long on this task or the overlay will begin to set and produce a "ripple" effect.

6. Allow the overlay to harden for several minutes.

7. Spot the compounds onto the center of the appropriate plates using the micropipet. Use 1 μL of the MMS and 10 μL of the other solutions per plate.

8. When the spot has dried, invert the plates and incubate at 37°C for 2–3 d.

9. Observe the plates, noting any inhibition of lawn growth or the presence or absence of a ring of revertant colonies.

4. Results and Discussion

Control plates should give the standard reversion frequencies for the three strains. For TA100, between 100 and 200 colonies should be seen; for T98, between 30 and 50; and for TA1537, between 5 and 10.

The presence of S-9 may slightly increase the above figures. If any large increase is seen, it suggests contamination of the S-9 mix and this should be tested for separately.

In all three MMS plate strains, a clear toxic zone will appear around the spot. For TA100 only, however, a ring of revertants should appear outside this zone. This is because MMS is a base substitution mutagen.

A small zone of inhibition will appear for all three 9-aminoacridine plate strains. TA1537 should also show a ring of revertants around the toxic area, since 9-aminoacridine is a frameshift mutagen with a specificity for the TA1537 type sequence.

The 2-aminoanthracene plates will show no apparent effect.

In the 2-aminoanthracene + S-9 plates, if the S-9 is active, revertants will appear close to the spot for all three strains.

The results in the test compound plates will, of course, depend on what you test.

References

1. Ames, B.N., McCann, J., and Yamasaki, E. (1975) Methods for detecting carcinogens and mutagens with the *Salmonella/* mammalian microsome mutagenicity test. *Mutation Res.* **31,** 347–364
2. Maron, D.M. and Ames, B.N. (1983) Revised methods for the *Salmonella* mutagenicity test. *Mutation Res.* **113,** 173–215

Appendix I

Glassware and Plasticware

Experiments in molecular biology require that all items of glassware and other equipment are scrupulously clean before use. This is particularly important for reactions involving proteins and nucleic acids carried out in small volumes, because dirty test tubes, bacterial contamination, or traces of detergent can easily inhibit reactions or accidentally degrade nucleic acids by the action of unwanted nucleases. Following thorough washing (dichromate–sulfuric acid solution is very effective for cleaning pipets), glass or plastic vessels or pipets should be thoroughly rinsed in distilled or double-distilled water and sterilized by autoclaving. Alternatively, heat-resistant glassware can be baked in an oven at 150°C for 1 h.

Glassware can be treated with 0.1% diethylpyrocarbonate overnight at 37°C as a precaution against ribonuclease contamination. The diethylpyrocarbonate is removed by heating the drained glassware at 100°C for 15 min. This procedure is not normally required for sterile disposable plasticware.

Very small quantities of nucleic acids can be absorbed onto the surface of glassware. The problem is significantly reduced if small polypropylene (e.g., Eppendorf) test tubes are used, but if these are not available, you may wish to coat glassware with silicone. To do this, soak or rinse glassware in a 5% (v/v) solution of dichlorodimethylsilane in chloroform. This solution can be stored at room temperature in a fume hood virtually indefinitely and used as required. Following this silicone treatment, thoroughly rinse the glassware and bake at 180°C overnight.

Many of the experiments in this book involve the use of solutions containing phenol and/or chloroform. These

265

solutions can be successfully used in glass or polypropylene (Eppendorf) tubes, but attack many other forms of plastic, particularly polycarbonate, as used in many types of centrifuge tube. Check the manufacturer's recommendations if you are doubtful about using a particular type of plastic tube and remember that sealing films are also soluble in phenol and chloroform.

Preparation of Buffers and Solutions

When making up solutions, observe the following guidelines:

1. Use the highest grade of reagents, when possible.
2. Prepare all solutions with the highest-quality distilled water available.
3. Autoclave solutions when possible. A guide to this is given for each chapter in Appendix II. If the solution cannot be autoclaved and you wish to store it, sterilize by filtration through a 0.22-μm filter.
4. Check the pH meter carefully, using freshly prepared solutions of standard pH, before adjusting the pH of buffers.
5. Store solutions frozen when possible.

Many of the buffers and solutions in this book are used in more than one procedure or are used with minor alterations. A lot of time can be saved by making up concentrated stock solutions (e.g., $1M$ Tris) that can be used to make up a range of different solutions.

Phenol is a commonly used organic compound for the deproteinization of nucleic acid solutions. Most commercially obtained phenol preparations, particularly those that are even slightly discolored, contain oxidation products that are reported to cause breakdown or cross-linking of RNA and DNA. It is advisable therefore to redistill phenol at 160°C to remove contaminants. **Note**: *Care should be taken with this step because phenol is a particularly nasty chemical and causes severe skin irritation and burns. Wear safety glasses and gloves, and use a fume hood, electrical heating mantle, and boiling chips to reduce the possibility of an accident. Do not attempt to distill all the phenol from the distillation flask, since you will be over-concentrating some highly dangerous chemicals.*

Collect the distilled phenol in a vessel containing some water, which hydrates the phenol and lowers the melting temperature. Store the redistilled phenol at −20°C in aliquots in tightly capped, labeled containers, and melt as and when required.

Quantitation of Nucleic Acids

Nucleic acids in solution absorb UV light at a peak wavelength of 260 nm. The absorbance is, to an extent, dependent on whether the nucleic acids are single- or double-stranded, but a spectrophotometric measurement is reliable and accurate enough for most experiments. An absorbance (A, or optical density, od) at 260 nm of 1.0 in a 1-cm light path corresponds to a concentration of 50 μg/mL for double-stranded DNA and 40 μg/mL for single-stranded DNA and RNA.

It is advisable to scan the absorbance of nucleic acid solutions between 240 and 300 nm or to measure the absorbance at 260 and 280 nm. This is because proteins also absorb UV light (peak wavelength, 280 nm) and comparison of the absorbance at 260 and 280 nm gives an indication of the purity of the nucleic acid solutions. The A_{260}/A_{280} ratio should be 1.8 and 2.0 for DNA and RNA, respectively. Ratios lower than this indicate protein contamination and accurate measurements of the nucleic acid concentration will not be obtained.

If the nucleic acid concentration is too low to be accurately detected by spectrophotometry, an aliquot of the sample can be run on minigels alongside known standards (10–100 ng) of, for example, λ DNA, according to the procedure described in Chapter 9. Comparison of the staining intensity of the unknown with that of the standard will give an estimate of the unknown's concentration.

Quantitation of Proteins

A number of different methods are available for measuring the concentration of protein in solution and they all have their relative advantages and disadvantages. Three methods are given below. The simple absorbance method and Biuret test are quick but rather insensitive. The Lowry method is sensitive and reproducible, but requires the construction of a calibration curve.

Absorbance

Proteins in solution absorb UV light at a peak wavelength of 280 nm. An absorbance (A, or optical density od) at 280 nm of 1.0 in a 1-cm light path corresponds to a concentration of 1.8 mg/mL, dependent on the tyrosine and tryptophan content. A correction can be made for contaminating nucleic acids by measuring the absorbance at 280 and 260 nm and applying the following formula:

Protein concentration (mg/mL) =

$$(1.5 \times A_{280}) - (0.75 \times A_{260})$$

When measuring the absorbance, always use a blank with the same buffer composition as that containing the sample.

Biuret Test

To make up the Biuret reagent, dissolve in order: 2.25 g sodium potassium tartrate, 0.75 g copper sulfate·$5H_2O$, and 1.25 g potassium iodide in 100 mL of $0.2M$ NaOH. Make up to 250 mL with distilled water and store in a plastic bottle.

Mix 0.9 mL of this reagent with 0.1 mL of sample and leave at room temperature for 20 min. Read the absorbance of the solution at 550 nm against a Biuret reagent and buffer blank (note that NH_4^+ ions interfere with the assay).

An absorbance at 550 nm of 1.0 in a 1 cm light path corresponds to a protein concentration of approximately 3.3 mg/mL.

Lowry (Folin-Ciolcalteau) Method

This method is based on the work of Lowry et al. (J. Biol. Chem. **193**, 265, 1951). The reaction is based upon the ability of tyrosine residues in a protein treated with Cu^{2+} in alkaline solution to reduce a mixture of phosphomolybdate and phosphotungstate salts to give a blue color. It is a more sensitive method than the Biuret method—as little as 5 μg protein can be detected—but since it relies on the tyrosine content of the protein, the use of one specific protein as standard to measure the concentration of another protein is incorrect. Even so, because of its convenience and sensitivity, it is widely used.

Reagents

2% Sodium carbonate in 0.1N sodium hydroxide (A)
1% Sodium potassium tartrate (B)
0.5% Copper sulfate (C)
1.0N Folin reagent
Standard egg or serum protein solution (0.1 mg/mL)

Mix solutions A, B, and C in the proportion 50:1:1 to make reagent D immediately before use.

Procedure

Construct a calibration curve using between 0 and 0.1 mg protein, add 5 mL reagent D, and adjust the total volume to 6.0 mL. Allow to stand at room temperature for 10 min, then add 0.5 mL $1.0N$ Folin reagent and allow to stand for a further 30 min. Determine the absorbance at 660 nm using an appropriate blank. Treat the unknown solution in a similar manner.

Safe Handling of Microorganisms

Several of the protocols described in this book, and many procedures used in molecular biology research, involve the use of live microorganisms. Whenever such organisms are used, it is essential that permission is sought from the departmental safety officer and that laboratory workers adhere rigidly to a microbiology laboratory code of practice and thereby significantly reduce the possibility of causing a laboratory-acquired infection.

The safest way to approach work with live microorganisms is to make the following assumptions:

1. Every microorganism used in the laboratory is potentially hazardous.
2. Every culture fluid contains potentially pathogenic organisms.
3. Every culture fluid contains potentially toxic substances.

The basis of a microbiology laboratory code of practice is that no direct contact should be made with the experimental organisms or culture fluids, e.g., contact with the skin, nose, eyes, or mouth. It must also be noted that a large proportion of laboratory-acquired infections result from the inhalation of infectious aerosols released during laboratory procedures. Below is a list of instructions that forms the basis of a microbiology laboratory code of practice.

1. A laboratory coat that covers the trunk to the neck must be worn at all times.

2. There must be no eating, drinking, or smoking in the laboratory.

3. There must be no licking of adhesive labels.

4. Touching the face, eyes, and so on should be avoided.

5. There must be no chewing or biting of pens or pencils.

6. Available bench space must be kept clear, clean, tidy, and free of inessential items such as books, handbags, and so on.

7. No materials should be removed from the laboratory without the express permission of the laboratory supervisor or safety officer.

8. All manipulations, such as by pipet or loop, should be performed in a manner likely to prevent the production of an aerosol of the contaminated material.

9. Pipeting by mouth of *any* liquid is strictly forbidden. Pipet fillers, teats, or automatic pipets are used instead.

10. All manipulations should be performed aseptically, using plugged, sterile pipets, and the contaminated pipets should be immediately sterilized by total immersion in a suitable disinfectant.

11. Contaminated glassware and discarded Petri dishes must be placed in lidded receptacles provided for their disposal.

12. All used microscope slides should be placed in receptacles containing disinfectant.

13. It must be recognized that certain procedures or equipment, e.g., agitation of fluids in flasks, produce aerosols of contaminated materials. Lids should be kept on contaminated vessels, when possible.

14. All accidents, including minor cuts, abrasions, and spills of culture fluids and reagents must be reported to the laboratory supervisor or safety officer.

15. Before leaving the bench, swab your working area with an appropriate disinfectant fluid.

16. Whenever you leave the laboratory, wash your hands with a germicidal soap and dry them with paper towels. Laboratory coats should be removed and stored for future use or laundered. Do not under any circumstances wander into an office or restroom area wearing a potentially contaminated laboratory coat.

Safe Handling of Radioactivity

The use of radioisotopes has made a highly significant impact on the field of biochemistry. The ability to detect radioactive isotopes in minute amounts both accurately and reproducibly has facilitated clinical diagnosis and the study of cell metabolism and gene expression. Contemporary molecular biology and gene cloning rely heavily on the use of radioisotopes and any training given in molecular biology must therefore include procedures that employ radioactive compounds. It must be remembered, however, that radioisotopes are potentially hazardous and their use requires the application of strict safety precautions. Permission must be obtained from the Radioisotope Safety Officer (Radiation Protection Advisor), or relevant authorities, before experiments involving radioisotopes can be carried out.

The majority of radioisotopes used in biological experiments (e.g., ^3H, ^{14}C, ^{35}S, and ^{32}P) emit ionizing radiation as β-particles (electrons). The energy of these β-particles depends on the source, and influences the penetrating power and potential hazard of the radiation. Tritium, ^3H, emits β-particles of very low energy, such that the radiation is incapable of traveling more than a few millimeters in air and is readily absorbed by plastic or glass containers. A consequence of this low penetration is that ^3H cannot be detected by conventional Geiger counters, which form the basis of most laboratory safety monitors. Carbon-14 and ^{35}S emit β-particles of higher energy than ^3H and can be detected by a Geiger counter. The radiation from these isotopes is also, however, readily absorbed by plastic or glass, thereby significantly reducing the hazard. Phosphorus-32 emits radiation of high energy and is potentially very hazardous. Safety glasses and perspex screens or lead bricks are required to shield the operator from the ^{32}P source.

Several of the experiments described in this book involve the use of radioisotopes. When relatively large amounts of ^{32}P are involved (e.g., Chapters 5 and 12), it is recommended that the initial radiolabeling procedures be carried out by a suitably qualified and registered demonstrator or technician in a fully equipped licensed laboratory. Following removal of excess isotope, low levels of ^{32}P and the weak β-emitters (^3H, ^{14}C, and ^{35}S) can be used by participants in the practical class without needing shielding screens or elaborate facilities. The use of low levels of radioactivity is good experience for students of molecular biology and involves little hazard provided that a number of basic laboratory safety procedures are observed.

The principal hazards concerned with low energy β-radiation emitters arise from possible ingestion or skin contamination. Below is a list of basic safety precautions designed to reduce risks in the radiobiology laboratory.

1. A laboratory coat that covers from the trunk to the neck must be worn at all times.
2. Rubber disposable gloves should be worn, monitored frequently, and discarded if contaminated.
3. There must be no eating, drinking, or smoking in the laboratory.
4. There must be no licking of adhesive labels.
5. Touching the face, eyes, and so on should be avoided.

6. Available bench space must be kept clear, clean, tidy, and free of inessential items. Briefcases, handbags, and so on should be left outside of the laboratory.

7. Pipeting by mouth of *any* liquid is strictly forbidden.

8. Contaminated, used glassware must be placed in receptacles containing decontaminating detergent solutions.

9. Contaminated solid waste, e.g., gloves or tissues, should be discarded in waste receptacles provided.

10. Liquid waste should be poured into receptacles provided, not down the sink.

11. All vessels containing radioactive materials should be clearly labeled.

12. All work should be done on a nonabsorbant surface covered in Benchcote or with a spill-tray.

13. Any spills or accidents should be immediately reported to the laboratory supervisor. In case of a spill, mop up as much of the liquid as possible with paper towels, but do not spread the contamination. Immediately seek the advice of the Radioisotope Safety Officer.

14. Before leaving the laboratory, monitor your working area, hands, and clothing. Wash your hands and dry with paper towels.

Autoradiography

Introduction

As its name suggests, autoradiography is the process by which a radioactive object produces an image of itself on photographic film. This is a very useful technique in molecular biology, being used, for example, to detect radioactive protein bands in polyacrylamide gels after electrophoresis of products of in vitro translation, for the identification of radioactive bands on a Southern blot after hybridization with a radioactive RNA or DNA probe, or for the identification of bacterial colonies containing DNA complementary to a radioactive probe.

Before describing the various methods available for autoradiography, it is worth giving a brief description of the photographic process. Photographic emulsion contains a densely packed layer of microscopic silver halide crystals. When a photon of light is absorbed by one of these crystals, one silver ion is converted into a silver atom, which has a half-life of about 1 s at room temperature, after which it reverts to the ionic state unless stabilized by a second silver atom. As a result of treatment with a developer, a crystal will either have all its silver halide converted to silver, in which case it appears black or will remain as silver halide, which appears white or colorless. The greater the number of silver atoms already present in a crystal, the greater its chances of being developed into a black grain; five silver atoms give a 50% chance of development. This is the basis of the photographic process: electromagnetic radiation forms a "latent" image by creating some silver atoms in the silver halide crystals, and subsequent development turns the "primed" crystals black.

Direct Autoradiography

There is sufficient energy in the radiation emitted by most common radioisotopes to produce several silver grains in any crystal irradiated, thus forming a stable latent image. It is therefore quite possible to place a sheet of untreated film in contact with the radioactive gel or filter to obtain a "direct" autoradiogram. However, most of the radiation of high-energy emitters (e.g., ^{32}P or ^{125}I) will pass through the film without being absorbed, and that of low-energy emitters will largely be absorbed by the gel or filter before it can reach the film. Clearly, direct autoradiography is an insensitive method that can only be used with high levels of radioactivity.

Indirect Autoradiography (Fluorography)

In this technique the radiation from the isotopes is absorbed by a fluorescent compound that re-emits the energy as light (normally UV). For high-energy emitters, the fluor is in the form of an intensifying screen (e.g., calcium tungstate) that is placed against the film on the opposite side to the radioactive source; any radiation that passes right through the film is then absorbed by the screen, where light is emitted back onto the film. In this way much more of the emitted radiation is used, and the film is exposed by a combination of direct and indirect autoradiography. Because the radiation spreads as it passes through the film, and again when re-emitted as light, there is a slight loss of definition in the image, but this is usually more than compensated for by the increased sensitivity. Intensifying screens are often omitted when autoradiographing sequencing gels, since the highest possible resolution is needed and samples are usually very "hot."

Low-energy emitters have the opposite problem, since their radiation is too readily absorbed. Consequently, the fluor is infiltrated into the gel or filter, where it can absorb the radiation close to its source, and re-emit light that will pass through the gel or filter to the film. The most common fluor for use with ^3H, ^{14}C, or ^{35}S is PPO (2,5-diphenyloxazole), although sodium salicylate can be used instead. Several proprietary liquid fluors are available, offering improvements in safety and ease of use compared with PPO or salicylate.

Reciprocity Failure

It might be expected that the density of the final image after development would be proportional to the total energy that had fallen on the film, since more energy produces more silver atoms, which increases the chances of a grain becoming black. However, the actual response is not linear. At very low levels of radiation, no image is formed, regardless of the length of exposure (reciprocity failure). This is a consequence of the instability of single silver atoms in a crystal: Unless a second silver atom is formed by another photon within the lifetime of the first atom, the silver will revert to its ionic form, and the crystal will no longer be "primed." Thus, a certain threshold level of radiation is needed before the film will give an image.

There are two ways to overcome this reciprocity failure:

1. Pre-exposure of the film to a controlled amount of light will result in the generation of two or more silver atoms per crystal; this is a stable state. Hence every extra photon absorbed counts toward the final image density. (For procedure, *see* the section entitled "Prefogging.")

2. At temperatures of $-70°C$, the lifetime of a single silver atom is prolonged, thus increasing the chances that a second atom will be generated before the first reverts.

In combination with fluorography, these methods will produce an image from less than 3000 dpm of ^3H in a band 0.1 x 1 cm, after only 24 h exposure.

Practical Procedure

Autoradiography of Tritium-Labeled Hybridization Filters

1. Place the dried filter in a glass Petri dish and wet it evenly with 0.7% (w/v) PPO (2,5-phenyloxazole) in ether. Do this in a fume cupboard and wear rubber gloves for protection against the carcinogenic PPO.

2. When the ether has evaporated, mark the filter with radioactive ink to allow identification and correct orientation of the final autoradiogram against the filter. It is essential that the developed autoradiogram can be correctly superimposed over the filter, so that radioactive colonies or bands can be picked out.

3. This stage must be performed in a darkroom, using a suitable safelight such as Kodak GBX. This transmits very little light, so it will take about 5 min for your eyes to become fully adjusted to the dim light, after which any gaps around doors or ventilators will become glaringly obvious. Place a dried, PPO-impregnated filter on one of a pair of thick Perspex plates and hold it in position using a single layer of Saran wrap. Position a sheet of prefogged (see below) X-ray film on top of the filter, and complete the sandwich with the second of the Perspex plates. Use a little self-adhesive tape to hold the assembly together, and then wrap it in a thick plastic bag (e.g., those supplied around photographic paper); tape the bag closed. When you have checked that all sheets of film are safely put away, turn on the room light, and label the fluorography package, which should then be placed in a freezer at $-70°C$.

   After a suitable amount of time, which will depend on the specific radioactivity of the probe, allow the package to reach room temperature, then unwrap *IN THE DARKROOM*. Do not tear off the inner tape too rapidly: This may generate sparks that could give artifactual images on the film.

   Develop the film in undiluted Kodak D-10 for 5 min, agitating gently for the first 10 s only. Wash the film quickly, then fix in any normal film fixer (e.g., Ilford "Hypam") for 1 min, until the creamy emulsion has cleared to leave a black image on a light gray background. After a thorough wash (about 30 min), the film can be hung up to dry.

Autoradiography of ^{32}P-Labeled Hybridization Filters

The procedure for autoradiography of ^{32}P-labeled hybridization filters is the same as for ^{3}H-labeled filters, but

impregnation of the filter with PPO is omitted. An intensifying screen, placed on the side of the film opposite to the
sample, will increase sensitivity up to tenfold. Note that
preflasing of film and exposure at $-70°C$ only increase sensitivity of detection of ^{32}P when intensifying screens are
used (Laskey, R.A. and Mills, A.D., 1977, *FEBS Lett.* **82**,
314–318).

Notes

Prefogging

The aim of prefogging is to expose the film to
sufficient light to generate between two and five silver
atoms per grain, thus taking the film past its "threshold"
exposure onto the linear part of the "image density" vs
"exposure" curve. The correct conditions for this prefogging must be found by trial and error, and should then be
reproducible. It is common to use an electronic flashgun
(screened by filters) or a photographic enlarger as the light
source, the intensity being varied by altering the distance
between light and film, and/or by changing the number of
filters, the lens aperture, or the length of exposure. A set of
strips cut from the film should be exposed to various light
intensities, and then developed and fixed as normal. The
absorbance of each strip should be measured at 540 nm,
using unexposed, but processed, film as blank. Choose prefogging conditions that given an absorbance of about 0.15.

Test of Darkroom

It is easy and useful to test if the darkroom and
safelight really are safe. With the safelight turned on, place
a strip of prefogged film on the work surface and cover part
of the film with a coin. After 10 min, develop and fix the
film as usual; if a clear disk is seen surrounded by a dark
gray background, you should check for stray light or an
inefficient safelight.

Using an Ultracentrifuge

Ultracentrifuges are an integral part of every biochemistry department and are essential items for many procedures involving purification or analysis of biological molecules or organelles. By spinning at very rapid speeds, an ultracentrifuge is capable of generating enormous forces that are capable of separating subcellular organelles and macromolecules according to size or density. Modern ultracentrifuges generate a force in excess of 100,000g, where g is the acceleration caused by the earth's gravitational pull, and speeds at the perimeter of the rotor are in excess of the velocity of sound. It is not surprising that such machines are complicated and expensive; they should be treated with respect and the manufacturer's handbook consulted before use.

Listed below are some general aspects relevant to the use of all makes of centrifuge.

1. Rotors are expensive and become corroded by certain frequently used chemicals, e.g., cesium chloride and ammonium sulfate. Rotors must be thoroughly cleaned and dried before and after use.

2. Rotors are spun with one or more pairs of balanced tubes oppositely placed or with three tubes at the corners of an equilateral triangle. Some manufacturers recommend that swinging bucket rotors should be spun with all the buckets in place, regardless of the number of tubes.

3. Tubes should be balanced to within 0.2 g. Do not guess; weigh accurately.

4. Tubes should be full to prevent collapse during centrifugation (unless thick-walled polycarbonate tubes are being used). Fill noncapped swinging bucket tubes to within a few millimeters of the rim and fill capped tubes completely. A grub screw in the lid facilitates filling via a Pasteur pipet and mineral oil can be floated on the surface of the tube contents to fill the tube.

5. If the buckets and spaces on the rotor are numbered, always fit the buckets to their corresponding slots.

6. Centrifuges spin the rotor in a high vacuum. Check that all O-rings and gaskets on rotors or buckets are in good condition and slightly greased (if recommended).

7. Check the oil level and water supply to the ultracentrifuge before commencing the run.

8. Fill in the log book with full details of the run and the operator.

The Plasmid pBR322

Plasmid pBR322 (*see* Fig. 1) is an example of a synthetic plasmid commonly used in gene cloning. It contains a relaxed origin of replication (i.e., independent of chromosomal control), two antibiotic resistance genes, to act as selection markers for cells carrying the plasmid, and single sites for a number of common restriction enzymes. The plasmid is small (2.6 Mdaltons, 4362 base pairs) and the base sequence of the entire genome is known (Sutcliffe, J.S., *Cold Spring Harb. Symp. Quant. Biol.* **43**, 77–90, 1979). The diagram below gives the structure of pBR322 showing the position of unique cleavage sites by restriction enzymes.

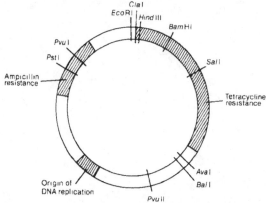

Appendix II

Details and Notes on Preparation

The following pages contain all the information required to carry out the experiments described in this book Chapter by Chapter. Preparation tasks required in advance are described where appropriate and lists given of the quantities of solutions, chemicals, and equipment required.

Chapter 1

General Preparation

The preparation details are for a class of 20 students. Larger numbers can, however, perform this exercise if laboratory space is available.

Cultures

1. *Plasmid carrying strain*—A culture of any strain carrying the plasmid pBR322 (or one of its derivatives) will be satisfactory, provided that it is not particularly resistant to lysis. We use JA221, which originated in the laboratory of Carbon, carrying pBR322. The culture is simply prepared by

growing overnight with gentle shaking at 37°C in LB broth. A 100-mL supply will be adequate for ten pairs of students.

2. *Recipient strain*—Any of the generally available strains selected for high transformation efficiency should be satisfactory for this exercise. We generally use BJ5183 in our classes. It is essential that the culture used in the experiment is in early log phase and growing rapidly. We achieve this by preparing an overnight culture (standing) at 37°C and using this to inoculate the cultures to be used for the class about 2 h before they are needed. We have found that 1 mL of the overnight culture inoculated into 100 mL of LB, followed by incubation at 37°C (shaking), produces satisfactory results. We will always, however, follow the growth be measuring A_{650}. A final reading of 0.2 is generally achieved. For ten pairs of students, two such flasks will be needed.

Solutions

Solutions given are for ten pairs of students. Sterilize by autoclaving, except where indicated.

1. 120 mL TE Buffer
    10 mM Tris, 1mM EDTA, adjust to pH 8.0
2. 5 mL 20% Sucrose in 5× TE
    20% Sucrose, 50mM Tris, 5 mM EDTA, adjust to pH 8.0
3. 2 mL 0.5M EDTA, pH 8.5
4. 2 mL Lysozyme
    10 mg/mL in sterile distilled water. Do not autoclave this preparation.
5. 10 mL Triton solution
    5% Triton X-100, 5 mM Tris, 50 mM EDTA, 8% sucrose, adjust to pH 8.0
6. 30 mL Absolute ethanol
    Precool in freezer (Do not autoclave!)
7. 30 mL 0.1M CaCl$_2$

Growth Media

1. 400 mL LB broth (including initial strain preparation)

   0.5% Yeast extract, 1% NaCl, 1% tryptone.

   Sterilize by autoclaving.

   Each pair will require 2 × 2 mL in small, sterile conical flasks.

2. 10 × 20 mL LB agar plates

   As LB broth plus 2% agar, added before autoclaving

3. 20 × 20 mL LB + ampicillin plates

   Prepare as LB plates, with ampicillin added after autoclaving to give 100 μg/mL. The ampicillin is prepared in sterile water at a concentration of 100 mg/mL and added at 1 mL/L of medium. It is sometimes necessary to add a small amount of NaOH to help the ampicillin dissolve.

4. 20 × 20 mL LB + tetracycline plates

   Prepare as LB plates, with tetracycline added after autoclaving to give 15 μg/mL. In this case the tetracycline is prepared in ethanol at 15 mg/mL.

Equipment

1. Bench centrifuges—ideally, a refrigerated bench centrifuge should be used. However, adequate results are obtained with normal room-temperature centrifuges if care is taken to keep all solutions cool.

2. Waterbaths—42°C (still), 37°C (shaking with accommodation for 20 small flasks).

3. Small glass bottles—two per pair, preferably thin-walled.

4. Sterile pipets (1 and 10 mL).

5. Ice buckets

Alternative Methods

One feature of this exercise is the exceptionally crude nature of the preparation. The DNA is prepared by spooling onto a rod to at least partially purify it from the other precipitated components. Therefore, this method positively discriminates *against* plasmid DNA that is less likely to be wound onto the rod than longer strands of chromosomal DNA. If more time were available, an extended version of this exercise could include a deproteinization step (such as the phenol/chloroform extraction procedure described in Chapter 2), followed by ethanol precipitation and collection by centrifugation of the total precipitate. Although this could improve yield and hence transformation frequency, it would add roughly 1 h to the practical class and make it less suitable for large classes.

Chapter 2

General Preparation

Growth of Cells

Cells are grown and harvested prior to the practical class by following the procedure described below.

Solutions and Equipment

1. 1.6 L LB broth—Yeast extract, 5 g; NaCl, 10 g; tryptone, 10 g; distilled water, 1 L. After autoclaving, add 10 mL 20% glucose (filter sterilized) to each liter of broth

2. 1.6 mL Ampicillin—Prepare stock of 50 mg/mL in sterile distilled water, using a little NaOH to dissolve the ampicillin initially. Add 1 mL of this stock to each liter of LB broth after the broth has been autoclaved to give a final concentration of 50 μg/mL

3. 1.5 mL Chloramphenicol—Prepare stock of 150 mg/mL in absolute ethanol. Add 1 mL to each liter of LB broth at the amplification step to give 150 μg/mL

4. 200 mL TES—Tris-base 30 mM; Na$_2$EDTA, 5 mM; NaCl, 50 mM; pH 8.0

5. 250 mL Tris/sucrose—Tris 50 mM; sucrose 25% (w/v); pH 8.0

287

6. Six 250-mL centrifuge bottles and six 50-mL centrifuge tubes will be needed for washing and harvesting the cells before the practical class.

Protocol

The following steps should be carried out at least two
days before the practical class begins.

1. Inoculate 100 mL of LB medium containing 50 μg
   ampicillin/mL with a loop of bacteria, and
   incubate overnight at 37°C. *Escherichia coli* JA221
   (pBR322) has been found to give good results, but
   this method should be successful with a wide
   range of strains.

2. The next morning, inoculate 1.5 L of prewarmed
   LB/ampicillin with 60 mL of the overnight culture, and incubate with shaking at 37°C. After
   about 1.5 h, remove a few mL of suspension and
   measure its absorbance at 660 nm, using LB
   medium as a blank. Repeat at intervals until the
   absorbance is about 0.4 (be prepared for a rapid
   increase in absorbance; doubling time may be as
   short as 30 min), then add 1.5 mL of a 15% (w/v)
   solution of chloramphenicol in ethanol, giving a
   final concentration of 150 μg/mL.

3. Incubate this culture, with shaking, at 37°C overnight (for at least 16 h; longer incubations will do
   no harm).

4. Harvest the cells by centrifugation at 2500g in six
   250 mL bottles at 4°C, for 10 min. (Note that all
   relative centrifugal forces are given as g average.)

5. Pour off the supernatants, which should be autoclaved before disposal. Resuspend the pellets in
   30 mL chilled TES buffer per bottle. This is most
   easily achieved by violent shaking of each (sealed)
   bottle, using a vortex mixer, until the pellet is
   resuspended in its own liquid to give a smooth
   paste; then the TES is added to give a homogeneous suspension, free of cell clumps.

6. Transfer the suspensions into six 50-mL centrifuge tubes, and centrifuge at 3000g for 10 min at
   4°C.

7. Decant supernatants and resuspend each pellet in 30 mL of chilled Tris/sucrose. It is usually convenient to place these suspensions in a freezer, where they can be stored indefinitely for future use. Although they can be used immediately, freezing does seem to help subsequent lysis of the cells, so it is worth including a freezing/thawing cycle at this stage.

8. Immediately before the practical class starts, thaw the suspensions and centrifuge at 5000g for 10 min at 4°C to harvest the cells. Give one centrifuge tube to each pair of students, allowing them to decant the supernatant and resuspend the bacterial pellet.

Requirements for Plasmid Isolation

It is likely that the maximum number of students that can be accommodated will be determined by the CsCl ultracentrifugation. Quantities of reagents have been calculated on the assumption that a six-place swing-out rotor will be available, allowing a maximum of six pairs of students.

There is no need to autoclave these solutions as long as they are prepared in clean glassware with freshly distilled water, and are used immediately or after brief storage at 4°C.

Solutions and Chemicals

Volumes given are for six pairs of students.
1. 25 mL Tris/sucrose—Tris 50 mM, sucrose 25% (w/v), pH 8.0
2. 6 mL Lysozyme, 5 mg/mL in 0.25M Tris, pH 8.0. Prepare immediately before use
3. 10 mL EDTA 0.25M, pH 8.0
4. 10 mL Sodium dodecyl sulphate (SDS), 10% (w/v)
5. 10 mL Tris 0.25M, pH 8.0
6. 10 mL Tris-base 2.0M, no pH adjustment
7. 100 mL Phenol/chloroform mixture, 1:1 (v/v)

8. 20 mL Potassium acetate 4.4M, autoclaved
9. 150 mL Absolute ethanol, stored at $-20°C$
10. 10 L SSC (Saline sodium citrate)—NaCl 0.15M; Sodium citrate 0.015M. This will be used as a tenfold dilution (0.1 $\times$ SSC), for which the volume is given
11. 23.4 g CsCl, six lots of 3.9 g
12. 3 mL Ethidium bromide 5 mg/mL
13. 50 mL Isoamyl alcohol
14. 30 mL Ethanol, 70% (w/v)
15. 1 Centrifuge tube, containing cells, per pair of students

Equipment

Assume that one of each item per pair of students will be needed, unless indicated otherwise.

1. Ice bucket
2. Vortex mixer
3. 2 Conical flasks, 100 mL
4. Water bath, 37°C (One per class)
5. Thick-walled polycarbonate centrifuge tube, 25- or 35-mL in size
6. 2 Measuring cylinders, 10 mL
7. 2 Thick walled glass centrifuge tubes, 15 mL
8. 2 Plugged Pasteur pipets
9. 2 Polypropylene centrifuge tube, 50 mL, plus cap
10. High-speed centrifuge and angle rotor for at least six 50-mL tubes (One per class)
11. Vacuum desiccator, or nitrogen gas cylinder (One per class)
12. Disposable gloves
13. Safety glasses
14. Ultracentrifuge tube, 5 mL, UV transparent. Beware of tubes made of polycarbonate containing an additive to protect them from damage by UV light; such additives can work so well that the tubes completely block UV
15. Ultracentrifuge and swing-out rotor to take six tubes (One per class)

16. UV lamp (long wavelength) (One per class)
17. Syringe, 1 mL, and wide-bore needle
18. 1.5-mL microfuge tubes
19. Dialysis tubing, pretreated with EDTA

Chapter 3

General Preparation

Solutions and Chemicals

The amounts given here are for 10 pairs of students. (Note that all containers and stoppers should be washed thoroughly to remove any traces of detergent.)

1. 2 Bottles of alcohol for washing casting plates, in wash bottles. Industrial methylated spirits is satisfactory.

2. 1 L Gel-running buffer × 10—Tris $0.9M$; Na_2EDTA 25mM; boric acid $0.9M$; adjusted to pH 8.2 using HCl. This stock solution is 10 times its final working concentration, and can be stored indefinitely at 4°C.

3. 10 × 100 mL Agarose solution—agarose (e.g., Type I, Sigma, or any equivalent preparation with a low sulfate content, giving a low coefficient of electroendosmosis), 0.8% (w/v) in single-strength gel-running buffer. Autoclave, or use a microwave oven, to dissolve the agarose; then keep at 50°C in a stoppered vessel until needed.

4. 250 μL DNA from bacteriophage λ, 2 A_{260} units/mL

5. 150 μL Plasmid pBR322, 0.5 A_{260} units/mL

6. 60 μL Restriction endonuclease *Eco*RI, 5 units/μL

7. 100 μL Restriction endonuclease *Hin*dIII, 5 units/μL

292

8. 5 ×1 mL Restriction buffer—Tris-HCl 1.0M; NaCl 0.5M; MgCl$_2$ 0.1M; pH 7.5. This is designed for use with *Eco*RI, but also gives satisfactory results with *Hind*III. Autoclave and freeze, to store.

9. 5 × 1 mL Sterile distilled water

10. 5 × 1 mL Gel-loading solution—Ficoll (type 400) 30%; bromophenol blue 0.25% (both w/v) in single-strength gel-running buffer. Store at room temperature

11. 5 × 1 mL Ethidium bromide, 5 mg/mL in single-strength gel-running buffer. *Wear disposable gloves when handling this compound; it is mutagenic.*

Equipment

Assume one of each item per pair of students will be needed, unless indicated otherwise.

1. Glass casting plate, 23 × 12.7 cm, 3 mm thick, with its edges smoothed. Perspex plates are not suitable, since they tend to warp

2. Zinc oxide tape, 12.5 mm wide

3. Scissors

4. Leveling table and spirit level (optional)

5. Well-forming comb—cut from Perspex, 3 mm thick, giving 10 teeth, each 9 mm across, separated by 2.8 mm gaps. This is glued to supports at each end so that when placed across the casting plate, the teeth are about 1 mm above the plate. The teeth must not be in contact with the plate or bottomless wells will result.

6. 6 Microcentrifuge tubes, 0.5 mL, polypropylene

7. Rack for microcentrifuge tubes

8. Adjustable, automatic micropipet, covering 2–20 µL

9. Disposable tips for micropipet, sterilized by autoclaving. (Approximately 40 tips per tray, five trays per class)

10. Microcentrifuge, with rotor to take 0.5 mL tubes. (At least one per class)

11. Water bath, 37°C, with support for microcentrifuge racks. (At least one per class)

12. Electrophoresis tank—made from Perspex, this
should be just wide enough to take the casting
plate plus two layers of zinc oxide tape. To avoid
any danger of the running buffer becoming
exhausted during electrophoresis, the tank is
designed to hold a large volume of buffer, but has
a relatively shallow central section where the gel
sits. Fig. 1 in Chapter 3 shows a typical tank. It is
advisable to ensure that power can be supplied to
the tank only when its lid is in place, even though
the voltages used are usually low.

13. Power supply—capable of producing direct
currents up to about 100 mA, and constant voltages up to 100 V. The output should be "floating" (i.e., not grounded), and must be protected
against short circuits. (Five dual output units per
class)

14. Shallow tray or plastic box, for staining gel. This
must be sufficiently large and shallow for easy
access to the stained gel

15. Pair of disposable gloves

16. Roll of plastic wrap (optional)

17. Transilluminator—ideally with its peak of emis-
sion at 300 nm, for high sensitivity with minimal
damage to the DNA. For protection against sunburn, the transilluminator should be screened by
a perspex screen, 5 mm thick. Safety glasses must
also be worn to protect the eyes; ordinary glasses
are not sufficient

18. Camera—a Polaroid camera is strongly recommended for photography of gels. The prints are
available immediately and are large enough for
easy interpretation and accurate measurements of
mobilities. The camera should be fitted with an
orange/red Wratten 23A filter to cut out any blue
light from the transilluminator, above a perspex
or glass filter to block UV light (the Wratten filter
fluoresces in UV light). Type 667 film is highly
sensitive and gives good-quality prints; type 665

is slower, but produces both negatives and prints. Good results can be obtained using 35 mm (or larger format) cameras with a film such as Ilford FP4 and the same filters as above, but the results are not available immediately.

Chapter 4

General Preparation

Amounts given are for 10 pairs of students.

Solutions

1. 5 mL Distilled deionized water
2. 1 mL 10× Concentrated restriction enzyme buffer:
   200 mM Tris-HCl, pH 7.5
   1M NaCl
   100 mM MgCl$_2$
   100 mM Dithiothreitol (DTT)
   100 μg/mL Gelatin

   The individual components (except DTT) should be autoclaved before mixing. This solution should be made prior to the practical class and stored at $-20°C$.
3. 0.3 mL pBR322 DNA (0.1 mg/mL)
   This can be obtained from most companies dealing with molecular biology reagents. Dilute with 10 mM Tris-HCl, pH 7.4, 0.1 mM EDTA to 0.1 mg/mL, and store at 4°C.
4. 0.3 mL λ DNA (0.1 mg/mL)
   Obtained, diluted and stored, as for pBR322.
5. 100 μL (100 units) EcoRI (1 unit μL)
   The enzyme is usually obtained in more concentrated form. It can be diluted according to the

manufacturers' instruction, but this should *only* be done immediately prior to use (i.e., the morning of the practical class), and then kept at $-20°C$ until required. New dilution should be done each day.

6. 100 μL (100 units) *Bam*Hl (1 unit/μL)
    See note in section entitled "Preparation."
7. 100 μL (100 units) *Sal* 1 (1 unit/μL)
    See note in section entitled "Preparation."
8. 2 mL Agarose gel-loading buffer:
    >20% Sucrose
    >10% Ficoll
    >10 m*M* EDTA
    >1% Bromophenol blue
    >Water

    Make up before practical class. This can be kept at room temperature.
9. 10 g Agarose
    Type II from Sigma, or equivalent.
10. 3 L 10 × Concentrated electrophoresis buffer:
    >0.9*M* Tris base
    >0.9*M* Boric acid
    >25 m*M* EDTA
    >pH 8.2, With hydrochloric acid
11. 30 L Deionized water for electrophoresis and preparation of agarose gel
12. 1 mL 10 mg/mL Ethidium bromide
    Make up in water. Always wear gloves when handling.

Apparatus Required

1. 10 Gel-forming assemblies and slot formers with tape
2. 10 Electrophoresis tanks and power packs
3. UV light box or transilluminator and facilities for photographing the gel (*see* Chapter 3)
4. 650 Pipet tips
5. 150 Microfuge tubes (1–5 mL) and racks
6. Bunsen burners and stands for boiling agarose

7. 10 Large trays for staining gels
8. Plastic face masks
9. Waterbaths, 37°C and 70°C
10. Semilog graph paper

Preparation

It is advisable to check that the DNAs are digested with the particular batch of restriction enzymes and with the dilutions that are suggested in Table 1 of Chapter 4. If digestion of either pBR322 or of λ is incomplete, the restriction enzyme should be used at a higher concentration.

For convenience you may wish to provide batches of agarose solution already melted and ready for pouring gels. In this case, carry out step 1 of the protocol and store the agarose batches in a 50°C water bath until required.

Chapter 5

General Preparation

Electrophoresis

The experiment described in this chapter uses an alternative design of electrophoresis apparatus (*see* Fig. 3b in Chapter 5) to that described in Chapter 3. This alternative is optional and does not affect the experimental results. If submerged gels are preferred, simply provide the equipment and gel-running buffer described in Chapter 3.

1. 5 g Agarose (Sigma Type II or equivalent)
2. 1.5 L 10 × Concentrated electrophoresis buffer:
   500 mM Tris base
   200 mM Sodium acetate
   20 mM EDTA
   pH 7.8 with glacial acetic acid
3. 10 L Deionized water
4. 100 μL (10 μg) *Eco*RI digested λ-DNA (0.1 mg/mL)
   10 μ λ-DNA are digested with 2 μL *Eco*RI (10 units/μL) in 100 μL of *Eco*RI digestion buffer (20 mM Tris-HCl pH 7.5; 10 mM MgCl$_2$; 200 mM NaCl) for 1 h at 37°C. The enzyme can then be heat-inactivated at 68°C for 10 min and the digested DNA stored at 4°C until required.

5. 500 μL Agarose gel-loading buffer
   20% Sucrose
   10% Ficoll
   10 mM EDTA
   1% Bromophenol blue
   H_2O
6. 1 L Ethidium bromide (1 μg/mL in water)
7. 10 Holders and combs for gels
8. 20 Buffer tanks (15 $\times$ 8 cm, 5 cm deep)
9. 10 Power packs and electrodes
10. Paper tissues for wicks
11. Trays for staining and soaking gels

Southern Blotting

1. 2 L Denaturation solution
   (0.5M NaOH, 1M NaCl)
2. 2 L Neutralization solution
   (0.5M Tris-HCl, pH 7.5, 3M NaCl)
3. 4 L 20$\times$ SSC
   (3M NaCl, 0.3M sodium citrate, pH 7.0)
   This solution can be diluted with water to give
   2$\times$ SSC, for wetting the filter
4. 10 Sheets nitrocellulose
   (same size as glass plate used for making gel)
5. 30 Pieces Whatman 3MM filter paper (15 $\times$ 20
   cm); 10 pieces Whatman 3MM filter paper (10 $\times$
   15 cm)
6. Paper towels or disposable diapers
7. Saran wrap
8. Tweezers
9. 80°C oven (preferably a vacuum oven)

Hybridization and Washing

1. Ten plastic bags
2. One heat sealer

3. 80 mL Prehybridization solution

      5× SSC (20 mL of 20× SSC)
      0.1 g Bovine serum albumin
      0.1 g Polyvinylpyrrolidone
      0.1 g Ficoll (mol wt 400,000)
      Water (19.2 mL)
      50% Formamide (40 mL)
      100 μg/mL sonicated, denatured salmon
        sperm DNA (0.8 mL of 10 mg/mL)

Add the ingredients in the order written. The solution may be made up in advance and stored at 4°C. To make the carrier DNA, dissolve salmon sperm DNA in 10 mM Tris-HCl, pH 7.5, 0.5 mM EDTA at 10 mg/mL, and sonicate until the viscosity is reduced markedly. Alter the solution to give 0.1M NaCl and boil for 5 min. Immediately place in ice and store at −20°C until required.

4. 80 mL Hybridization solution:

      3× SSC (12 mL of 20× SSC)
      0.02 g Bovine serum albumin
      0.02 g Polyvinylpyrrolidone
      0.02 g Ficoll (mol wt 400,000)
      Water (27.6 mL)
      50% Formamide (40 mL)
      50 μg/mL sonicated, denatured salmon
        sperm-DNA (0.4 mL of 10 mg/mL)

Again this can be made up in advance and stored at 4°C. When required, add heat-denatured radioactive probe to give a concentration of 10^6 cpm/mL. The probe is denatured by heating for 4 min in a microfuge tube placed in a boiling water bath. (Make small holes in the cap of the tube to prevent it from blowing open.) Add immediately to the hybridization solution and shake well.

5. 20 mL 20% Sodium dodecylsulfate (SDS)
      (20 g SDS/100 mL water)

6. 4L 2× SSC including 0.1% SDS

7. 2 L 0.1× SSC
      This can be prewarmed to 65°C before the class on day 3.

8. Ten film cassettes or holders
    If these are not available the filter and film may
    be positioned between two pieces of glass or
    hard plastic. This can then be held tight by tape
    and the package placed inside a black
    polythene bag.
9. Film for ten filters, for autoradiography
10. 42°C and 65°C waterbaths

Nick Translation

The volumes shown are for ten pairs of students,
though it is likely that this part of the procedure will be
demonstrated by a staff member.

1. If a kit is not available, make up the following:
    (i) 100 μL Nucleotide buffer solution (10× con-
    centrated)
    1 mM dATP
    1 mM dTTP
    1 mM dGTP
    0.5M Tris-HCl (pH 7.2)
    0.1M MgSO$_4$
    1 mM Dithiothreitol
    500 μg/mL Bovine serum albumen (DNAse-
    free grade)
    (ii) 10 μL DNAse-I solution (0.1 mg/mL stock
    solution)
    This should be diluted to 0.1 μg/mL in 10
    mM Tris pH 7.0, 50% glycerol just prior to
    use.
    (iii) 10 μL DNA polymerase-I (5 units/μL)
2. 2 mL Water-saturated phenol
3. 20 μL Sonicated salmon sperm carrier-DNA
    Dissolve the salmon sperm-DNA at 10 mg/mL
    in 10 mM Tris, pH 7.5, 0.5 mM EDTA, and
    sonicate until the viscosity is greatly reduced.
4. 30 mL G75 Sephadex
    Mix 10 g of Sephadex powder with 200 mL 10
    mM Tris pH 7.5, 0.1 mM EDTA. Allow to swell
    by leaving overnight or autoclaving. This ratio

of Sephadex powder to buffer is suitable for
pouring into the column after shaking.

5. Glass wool

6. Ten Pasteur pipets, tubing, and clips

7. Ten sets of 12 tubes and tube racks

8. 100 mL TNE buffer

> 10 mM Tris-HCl, pH 7.5
> 0.1 mM EDTA
> 0.1M NaCl

9. Equipment for working with radioactivity (*see*
Appendix I, "Safe Handling of Radioactivity.")

> Geiger counter
> Perspex shield
> Gloves and safety glasses
> Radioactive waste bin for solids and liquids
> Scintillant and vials (10)
> Scintillation counter

10. 500 μL α^{32}PdCTP (300 Ci/mmol, 1 mCi/mL) if
not provided in kit.

Note: ^{3}HdCTP may be used to make a labeled probe.
The advantage of this is that the probe, which is quite
expensive to make, can be made in bulk and stored for
many experiments. The disadvantage is that a scintillation
counter is required to monitor purification of the probe and
fluorography has to be used to detect the radioactivity on
the nitrocellulose and there is a long delay in obtaining
results. If a ^{3}H-labeled probe is to be used, refer to Appendix I, "Autoradiography," for details of the fluorography
procedure.

Chapter 6

General Preparation

Amounts given are for ten pairs of students.

Solutions

1. Deionized distilled water: 2 mL for diluting enzyme digests; 20 L, for electrophoresis
2. 100 μL 10 $\times$ concentrated EcoRI digestion buffer:
   200 mM Tris-HCl, pH 7.5
   100 mM $MgCl_2$
   1M NaCl
   1 mg/mL gelatin

   The individual components should be autoclaved before mixing. The solution may then be stored at 4°C.
3. 250 μL (50μg) λ-DNA (0.2 mg/mL)
   Dilute more concentrated solutions with 10 mM Tris-HCl, pH 7.5
4. 10 μL EcoRI (10 units μL)
5. 1 L 10 $\times$ Concentrated electrophoresis buffer:
   500 mM Tris base
   200 mM Sodium acetate
   20 mM EDTA, pH 7.8 with glacial acetic acid
   (Tris-borate buffer as described in Chapter 3 can be used for submerged gels)

6. 0.6 mL Agarose gel-loading buffer:
   20% Sucrose
   10% Ficoll
   10 mM EDTA
   1% Bromophenol blue
   Water

   This can be stored at room temperature.

7. 2 L Ethidium bromide solution (1 μg/mL)
   Make up in water. *Always wear gloves when handling.*

8. 40 mL Gel extraction buffer:
   50 mM Tris-HCl, pH 7.4
   10 mM NaCl
   10 mM EDTA

9. 40 mL Water-saturated phenol
   Solid phenol is melted at 60°C and shaken with an equal vol of 0.1M Tris-HCl, pH 8.0. The phenol will absorb about 20% of its own volume of water and the Tris buffer will neutralize the pH. The phases should be allowed to separate and the aqueous layer removed. The phenol is then shaken again with an equal volume of 0.1M Tris, pH 8.0. Most of the aqueous layer is again removed and the phenol can now be stored at -20°C until required. Phenol will oxidize in air to yield a variety of derivatives that can react with DNA. Normally phenol is redistilled before use and then stored under nitrogen, frozen. This is not necessary for this experiment, however, providing the phenol has no more than a faint coloration. Phenol can cause serious burns if contact with skin is made. Any areas of skin that come into contact with phenol should be washed extensively with soap and water. *Always wear gloves and safety glasses when handling and dispensing phenol.*

10. 5 mL Chloroform

11. 200 μL 5M NaCl

12. 20 mL Absolute ethanol

13. 1 mL TE buffer
     10 mM Tris HCl, pH 7.5
     0.1 mM EDTA

Materials

    5 g Low-gelling-temperature agarose
    5 g Agarose type II Sigma or equivalent
    10 Gel-forming assemblies and slot formers
    Electrophoresis equipment (as in Chapter 5 or submerged
        minigel apparatus)
    Ten buffer tanks
    Ten pairs of electrodes
    Ten power packs
    UV light box or transilluminator and photography equipment
    Face mask or safety glasses
    Razor blades

Preparation

Low-gelling-temperature agarose is rather expensive. Since each pair of students requires only two wells of a gel, it is possible, if necessary, to use a narrower plate on which to form the gel (say 3 or 4 slots wide), or for all the students to share one or two normal-sized gels.

The buffers and water-saturated phenol should be prepared before the class. The latter can be stored at −20°C until required. It is always worth the time taken to check that the enzyme digest conditions actually work with the enzyme preparation available (in this case EcoRI) before the practical class. The amount of enzyme can then be increased if digestion is not complete.

Furthermore, it would be worthwhile to have extra samples of EcoRI digested λ-DNA prepared in case some digestions fail during the class. This reserve can then be used for the electrophoresis steps.

Chapter 7

General Preparation

Amounts given are for ten pairs of students.

Solutions

1. 500 mL L broth:
   Difco Bacto Tryptone, 5 g
   Difco Bacto yeast extract, 2.5 g
   NaCl, 2.5 g
   Glucose, 0.5 g
   Mix, make up to 500 mL with water, and autoclave.

2. 40 mL Overnight stationary culture of HB101:
   Inoculate 40 mL of L broth with HB101 cells the day before the experiment. Shake overnight at 37°C. It is best not to use old cells when making competent cells, since this seems to decrease the transforming efficiency.

3. 0.5 mL Deionized distilled water for enzyme digestions and ligations

4. 50 μL 10 × Concentrated EcoRI digestion buffer:
   200 mM Tris-HCl, pH 7.4
   100 mM MgCl$_2$
   100 μg/mL Gelatin
   1M NaCl

307

The solutions should be autocalved separately in a more concentrated form and then mixed to give the correct concentration.

5. 50 μL pBR322 (0.2 mg/mL)
Dilute more concentrated solutions in 10 mM Tris-HCl, pH 7.5, 0.1 mM EDTA.

6. 10 μL *Eco*RI (2 units/μL)
If the enzyme is obtained in a more concentrated form, it should be diluted only immediately prior to its use, as recommended by the manufacturers. It is always worth checking that the enzyme will completely digest the DNA under the conditions described here before the class. If there is any evidence that the reaction is incomplete, then the concentration of enzyme used in the practical class may be increased.

7. 30 μL 10 × Concentrated ligase buffer:
   200 mM Tris-HCl, pH 7.5
   100 mM $MgCl_2$
   100 mM Dithiothreitol (DTT)
   2 mM ATP

8. 10 μL T4 DNA ligase (2 units/μL)
If the enzyme requires dilution, follow the manufacturer's instructions. Dilute only immediately prior to use.

9. 300 mL 0.1M $CaCl_2$
Sterilize by autoclaving.

10. 70 Ampicillin plates:
Make up L-agar as follows:
   L broth (*see* step 1 in this section)
   15 g/L Difco agar

Autoclave, cool to 55°C, and add 1 mL of a 0.1 mg/mL solution of ampicillin in water for every liter of L-agar. Mix gently and pour into Petri dishes (approximately 30–40 mL/plate). When solidified, these should be placed in a 40–50°C oven with the lids slightly open for 1–2 h to dry. They may be stored at 4°C for up to 2 wk.

Equipment

1. 37°C shaker incubator
2. 37°C oven
3. 37 and 68°C water baths
4. Benchtop centrifuge to take 50 mL tubes, in cold room, or a refrigerated bench centrifuge
5. Spectrophotometer and cells to read cell density
6. Pasteur pipets and Bunsen burner to make plate spreaders
7. 10 Sterile 250-mL conical flasks with cotton wool bungs
8. 20 Sterile 50-mL centrifuge tubes
9. 50 Sterile test tubes

Chapter 8

General Preparation

Preparation of Filters and Colonies

Before use, filters must be wetted and autoclaved. Millipore HATF filters (82 mm) are floated on distilled water. When the filters are uniformly wet, they may be submerged. The filters are then sandwiched between dry Whatman 3MM filters, wrapped in aluminum foil, and autoclaved.

The medium used in this exercise is LB (0.5% yeast extract, 1% NaCl, 1% tryptone, 2% agar). Fresh, still-moist LB plates are taken and a sterile Millipore filter laid on top of each one using alcohol flamed Millipore tweezers. When the entire filter is in contact with the medium, the filter is peeled off and inverted. Near an edge of the filter a needle is used to make a small hole for orientation of the filter. It may be convenient to make a pattern of small holes to enable individual filters to be recognized. One such filter and plate is prepared per pair of students.

Two bacterial strains are required for this exercise—one with and one without the plasmid pBR322. We use JA221 transformed with, or free of, pBR322, as in the chapter on microisolation of DNA (Chapter 9), but the choice of strains is not important. The strains are first grown overnight on LB plates, then picked onto the filter using a sterile needle. We have generally used a pattern consisting of four plasmid-carrying and five plasmid-free spots. The inoculated plates are grown at 37°C overnight before the practical class is scheduled.

Each pair of students will also require an LB plus chloramphenicol plate. Prepare a stock chloramphenicol solution of 150 mg/mL in absolute alcohol and add 1 mL to 1 L of medium. This will give a final concentration of 150 μg/mL.

Preparation of pBR322 Probe

The DNA probe is prepared by radiolabeling 1–2 μg of plasmid pBR322 (prepared according to the procedure described in Chapter 2, or purchased from a molecular biology products supplier) by a nick-translation reaction, prior to the practical class. This procedure is used to produce a ^{32}P- or ^{3}H-labeled probe with a specific activity of at least 10^{7} dpm/μg. Either isotope can be successfully used; ^{32}P has the advantage of giving rapid results on autoradiography (see Appendix I, "Autoradiography") but ^{3}H probes can be made in large quantities and stored. This is particularly useful if the class is to be run several times; it allows the probe to be conveniently tested, by going through the hybridization protocol several days before the laboratory class begins. The disadvantage of ^{3}H probes is that fluorography has to be used (as described in the protocol and Appendix I, "Autoradiography") and the X-ray film requires several days exposure.

All the reagents for nick translation are readily available from suppliers of molecular biology products. Some companies have ready-made kits for nick translation, and although these can be a little more expensive they are very useful for those unfamiliar with the technique and produce successful results with little or no previous experience. If the kit is not being used, follow the procedure for nick translation described in Chapter 5. If a ^{3}H probe is being prepared, a Geiger counter is not suitable for monitoring the eluent from the G75 Sephadex column (described in Chapter 5) used to remove unincorporated label; a scintillation counter is required. We have found, however, that column purification of probes is not necessary. Alcohol precipitation of the probe removes enough unincorporated label to allow successful detection of plasmid-carrying colonies described in this chapter. One microgram of high specific-activity labeled pBR322 is sufficient for ten colony hybridization filters. The labeled probe is dissolved in 1.5 mL sterile distilled water and denatured by incubation at 100°C in a boiling waterbath for 5 min immediately before use. Each filter for hybridization is incubated with 0.1 mL of this probe, as described in section 3.3 in Chapter 8.

Solutions

Amounts given are for ten pairs of students.

1.5 L 0.5M NaOH
1 L 1M Tris, pH 8.0
500 mL 1M Tris, 1.5M/NaCl, pH 8.0
10 mL Hybridization Buffer
   5× SET, 1× Denhardts, 0.1% SDS
   (5× SET-0.75M NaCl, 0.15M Tris, 5 mM EDTA, pH 8.0
   Denhardts-0.02% Ficoll, 0.02% polyvinylpyrrolidone, 0.02%
    BSA)
10 L Washing buffer
   2× SET, 0.5% SDS
5L 3 mM Tris base
15 mL 0.7% (w/v) PPO (2,5-diphenyloxazole) in ether
Radioactive ink: Indian ink with radioactive label added to give 10 μCi/mL. Use the label (^{3}H or ^{32}P) employed in the nick-translation reaction.

Equipment

10 Millipore tweezers
20 Sheets of Whatman 3MM paper
20 Plastic trays
Vacuum oven (if not available use a vacuum desiccator in an
   oven)
10 Petri dishes
Sealing tape
Plastic bags
10 large beakers (or deep glass dish) capable of accomodating
   filters
68°C Oven (it may be convenient to also have a 68°C waterbath for the buffer)
Autoradiography equipment

Chapter 9

General Preparation

It is assumed that six pairs of students will be performing this exercise. This number would require three eight-lane minigel apparatus and two 12-place microfuges. If more equipment were available, large numbers of students could be accommodated. It is possible to stagger use of this equipment to allow for more students, but this would then lengthen the time required.

Cultures

Three bacterial isolates are required, the first plasmid-free, the second carrying pBR322, and a third carrying a pBR322 derivative, including an insert. It is clearly desirable to use the same basic bacterial strain and limit the variation to the plasmid carried. We use the strain JA221 (which originates in the laboratory of Carbon) and obtain the plasmid-carrying isolates by transforming with plasmid DNA (as in Chapter 1). The choice of bacteria is immaterial, however, provided the strain used is a suitable host for transformation. For our inserted plasmid we use yIP5, which carries a small segment of a yeast chromosome. It must be emphasized, however, that any pBR322 derivative with an insert between 0.5 and 4 kb would be suitable.

The strains are presented to the students as three streaks on an LB plate (0.5% yeast extract, 1% NaCl, 1% tryptone, 2% agar). The labels A, B, and C are used in an

313

order known to the demonstrators but not to the students. The cultures are grown overnight at 37°C before use. As an alternative, students could analyze single colonies from a mixed spread plate. With this method, however, it is unlikely that each pair of students would get a sample of each type of clone.

Solutions

1 mL Triton mix
  5% Triton X-100, 5 mM Tris, 50 mM EDTA, 8% sucrose, pH 8.0 Sterilize by autoclaving
1 mL Lysozyme in Triton mix
  2 mg/mL lysosyme in above buffer
2 mL Isopropanol
  Freezer cooled
1 mL TE buffer
  10 mM Tris, 1 mM EDTA, pH 8.0 (autoclave)
Gel-running buffer, enough for three gel tanks (see manufacturer's instructions)
  0.09M Tris, 2.5 mM EDTA, 0.09M boric acid, pH 8.2
40 mL 0.8% w/v Agarose
  Add 0.8% Agarose to gel running buffer, briefly autoclave
300 mL 1 μg/mL Ethidium bromide
  Made up in gel-running buffer
15 μL pBR322
  pBR322 DNA is commercially available. Make up to 25 μg/mL in TE or other storage buffer
10 μL Loading mix
  20% Sucrose, 10% Ficoll, 10 mM EDTA, 1% bromophenol Blue

Equipment

21 Sterile 1.5-mL microfuge tubes
1 Roll zinc oxide adhesive tape
Two microfuges
Boiling waterbath—arrange with rack to take microfuge tubes
Sterile micropipet tips, micropipets
Three Minigel apparatus and power packs (It may be possible to obtain equipment with more than eight lanes. This will reduce the number of sets of equipment needed.)
-20°C Freezer

Transilluminator (Less satisfactory results can be obtained with a hand-held UV lamp held above the stained gel. Ideally, suitable photographic equipment should be available, but this is not essential.)
UV opaque glasses

Additional Points

There are two steps that seem to be particularly critical in getting a good result. First, it is important that the cells are completely suspended in the Triton mix. Second, it is important that, on adding the lysozyme, the tube is immediately put into the boiling waterbath (boiling, not just hot!) and left undisturbed for the full 3 min. Removal of the tubes to adjust the tape, and so on, during this time will reduce the quality of the final result.

It should be possible to treat the preparations successfully with restriction enzymes if this is desired, although a second alcohol precipitation may be required to remove traces of the detergent if the enzyme used is particularly sensitive. In this case the standards used would be linear restriction fragments, as in Chapter 3.

Chapter 10

General Preparation

The number of participants in this experiment is limited by the number of samples that can be accommodated in the ultracentrifuge step. Most fixed angle rotors have eight places that correspond to 16 students working in pairs. The following list will provide enough materials for eight pairs of students.

Growth of Plant Material

In the interest of economy and sensory satisfaction we use seeds from ripe melons. These should be cleaned, allowed to dry in air, and stored until required. Alternatively, seeds of melon, cucumber, or squash can be purchased from a seed supplier.

It is important to minimize the contribution of DNA from bacteria that grow within the seed coat and on the surface of the seedlings during germination. Approximately 500 seeds should be surface-sterilized by soaking in ethanol:water (1:1 by vol) for 10 min, followed by domestic bleach (2% in water, by vol) for 10 min and rinsing three times with sterile distilled water. Wherever possible, these manipulations should be carried out under a sterile hood. Do not contaminate the seeds by touching them with fingers or other non-sterile objects. The seeds should be germinated in autoclaved moist compost, vermiculite, or on sterile paper towels for about 1 wk at 20–25°C in the dark.

Solutions and Chemicals

Some of the solutions used in this experiment contain strong deproteinizing agents. Care must be exercised when weighing the reagents, measuring volumes, and so on. *Wear disposable gloves and safety glasses when handling the solutions, and do not mouth-pipet. If you splash reagents on your skin, wash thoroughly with water as soon as possible.* The cesium chloride used for centrifugation corrodes many metals. Do not use metal spatulas. Do not let crystals or solutions of cesium salts contact metal apparatus.

1. Homogenizing Medium

To about 50 mL distilled water, add:

5 mL Phenol/m-cresol/8/hydroxyquinoline solution (*see* the solution below)
1 g Triisopropylnaphthalene sulfonic acid (sodium salt)
6 g 4-Amino Salicylic acid (sodium salt)

Stir until dissolved. Make up to 100 mL with distilled water. This should be a sufficient amount of solution for one experiment. Provide in an automatic dispenser if possible.

2. Phenol/m-Cresol/8-Hydroxyquinoline Solution

300 g Phenol crystals
42 mL m-Cresol
0.3 g 8-Hydroxyquinoline
90 mL Distilled water

Mix the reagents together and stir until dissolved. This process is endothermic and can be hastened if the mixture is placed in a *warm* water bath.

Ideally the phenol crystals should be white and the m-cresol solution colorless. Normally, however, they contain oxidation products that make the phenol crystals pink and the m-cresol red-brown in color. For this experiment this does not matter much, but for some applications the phenol and m-cresol should be redistilled before use.

3.   Phenol – Chloroform

Take 200 mL of phenol/m-cresol/8-hydroxyquinoline solution (as above) and add 200 mL chloroform. Allow to stand and remove any water that rises to the top. Provide in an automatic dispenser if possible.

4.   Ribonuclease Buffer

0.15M Sodium chloride
0.015M Trisodium citrate, pH 7
Make up 100 cm^3 of solution and autoclave.

5.   DNA Buffer

50 mM Tris-HCl, pH 7.5
1 mM EDTA
Make up 100 mL solution and autoclave.

6.   Ribonuclease Stock Solution

Take a bottle containing 5 mg of commercial pancreatic ribonuclease A. Add 5 mL ribonuclease buffer and gently shake until dissolved. Heat enzyme solution at 80°C for 15 min and store frozen until required.

7.   Other Chemicals

500 mL Ethanol (96% ethanol is suitable)
200 mL 80% Ethanol (ethanol:distilled water, 4:1 by vol)
50 g of Analar cesium chloride.

Apparatus and Equipment

1.   Ultracentrifuge and Accessories

8 Small mortars and pestles
40 Glass centrifuge tubes (approximately 15 mL capacity)
   with caps and racks
8 Pairs of forceps
Pasteur pipets with rubber teats
8 Ice buckets

Bench centrifuges
A fine balance for weighing cesium chloride
Refractometer
Waterbaths set at 37°C

The experiment will monopolize an ultracentrifuge, rotor, and caps for 2 d. A swinging-buckets rotor can be used, but a fixed-angle rotor gives much better results. An eight-place fixed-angle rotor with a tube capacity of 10–12 mL capable of being centrifuged at 30,000 rpm (about 57,000–60,000g) is suitable. It is essential that the rotor is rated for use at this speed with solutions of density 1.7 g/mL. If in doubt, consult the ultracentrifuge handbook. Beckman Type 50, 65, and 75 rotors are suitable. It is safe to use an aluminum rotor if the procedure outlined in section 3 is followed. However, a *titanium rotor is preferable* because it will not corrode if cesium chloride should leak from the tubes during the run. Check that the tube caps are in working order and that polyallomer or other suitable ultracentrifuge tubes are available. (*See* Appendix I, "Using an Ultracentrifuge.")

2. Spectrophotometer and Accessories

The minimum requirement for this procedure is a spectrophotometer, capable of measuring in the UV range and 1 mL quartz glass semimicro cells. A scanning spectrophotometer with a pen-recorder, plus a flow-cell for analysis of the gradients, is an advantage but is not essential.

3. Equipment Concerned With the Fractionation and Analysis of the Density Gradients

One simple method of doing this requires a long, hollow needle 15 cm × 1–1.5 mm id, a retort stand plus clamps, a peristaltic pump plus microbore tubing, and 60 small test tubes or vials for each pair of students. Fractions are either collected manually (Fig. 2 in Chapter 10) and measured individually for refractive index or absorbance at 260 nm, or the tube contents are pumped directly through a UV flow-cell and the DNA peaks recorded by a pen-recorder.

Chapter 11

General Preparation

The following list will provide enough materials for 16 students working in pairs.

Preparation of Radioactive 25S and 18S rRNA for Hybridization

This is prepared by growing young seedlings in 5-^{3}H-uridine and extracting labeled RNA from roots.

Materials

1. Approximately 50 melon seeds
2. Two 250-mL conical flasks (autoclaved)
3. Domestic bleach solution
4. 200 mL Ethanol
5. 500 mL Autoclaved distilled water
6. 20 mL 2 mM Uridine solution (autoclaved)
7. Ribosome extraction buffer:

    250 mM Sucrose
    200 mM Tris-HCl pH 8.5
    50 mM KCl
    20 mM Mg(CH$_3$COO)$_2$

    Autoclave the solution and store in a refrigerator before use. Add 0.5% β-mercaptoethanol immediately before use.
8. 10% Sodium dodecyl sulfate (SDS) in water
9. "Phenol–Chloroform" mixture (50 mL) (*see* Chapter 10 in this appendix.)

10. 100 mL 3M Sodium acetate solution adjusted to pH 6 with acetic acid, and autoclaved

11. 50 mL 2 × SSC (0.3M NaCl, 0.03M Na citrate, pH 7.0) Autoclave the solution immediately after making it up.

Protocol

1. Surface-sterilize the seeds as described for Chapter 10 in this Appendix.

2. Germinate 30–50 seeds in a small sterile flask in a few mL sterile distilled water at about 25°C in subdued light. Inspect the seeds periodically and add a small volume of sterile distilled water when necessary. When the radicles begin to emerge, add high-specific radioactivity 5-^{3}H-uridine to the water in the flask, to a final concentration of 50–200 μCi/mL. Allow the seedlings to grow for 2 d, adding sterile distilled water when necessary. Take appropriate precautions for working with radioisotopes. (*See* Appendix I, "Safe Handling of Radioactivity.")

3. After 2 d, withdraw any liquid from the flask, add several mL of 2 mM unlabeled uridine solution and grow the seedlings for a further day.

4. Remove the seedlings with forceps, cut off the roots, and place them in a small chilled mortar. Freeze the tissue by adding liquid nitrogen and grind the roots to a fine powder while frozen. Add more liquid nitrogen, if necessary, to keep the tissue frozen.

5. Place 3 vol ribosome extraction buffer in a separate, chilled mortar and add the frozen, powdered root tissue. Homogenize the mixture with a pestle for a few minutes. Do not worry if the mixture freezes. Allow to thaw *slightly* and (wearing gloves) *squeeze* the mixture through a small piece of muslin previously moistened with extraction buffer. Centrifuge the filtrate at 13,000g for 30 min at 4°C, to pellet debris, plastids, and

mitochondria, and carefully pour the supernatant into a clean centrifuge tube.

6. Add sodium dodecyl sulfate to a final concentration of 0.5% (from 10% stock solution SDS), mix the contents of the tube, and add an equal volume of phenol–chloroform mixture (*see* Chapter 10 in this Appendix). Cap the centrifuge tube (do not use parafilm), shake to form an emulsion, and centrifuge the tube at 2000*g* for 15 min.

7. Withdraw the top aqueous layer carefully with a Pasteur pipet, taking care to leave the white protein layer behind, and transfer to a clean centrifuge tube. Add fresh phenol–chloroform, shake, and centrifuge at 2000*g* for 15 min as before.

8. Carefully pipet the supernatant into a clean centrifuge tube. Add 0.1 vol of 3*M* sodium acetate, pH 6, plus 2.5 vol of ethanol, thoroughly mix the contents, and store in a deep freeze (−15 to −20°C) for at least 2 h or overnight.

9. Centrifuge the tube at 2000*g* for 15 min to pellet the RNA. Drain the tube and add a small volume (2–4 mL) of cold, sterile 3*M* sodium acetate. Cap the tube and resuspend the contents by shaking. Centrifuge the suspension at 10,000*g* for 20 min at 4°C. The high-mol-wt RNA pellets on the bottom of the tube, whereas the DNA, 5S RNA, and tRNA remain dissolved.

10. Pour off the supernatant and drain the tube. Add 5 mL ethanol, cap the tube, and resuspend the pellet by shaking. Centrifuge at 10,000*g* at 4°C for 20 min.

11. Drain the tube and dissolve the RNA pellet in a small volume (1–4 mL) 2 × SSC (see below). Keep the solution cold. Withdraw a small aliquot and dilute to 1 mL. Measure the absorbance at 260 nm, using 2 × SSC as a blank, and calculate the RNA concentration (40 μg RNA dissolved in 1 mL has an absorbance of 1 in a 1-cm light path).

12. Withdraw another aliquot of RNA from the stock solution (10–20 μL) and spot this on to a nitrocel-

lulose filter. Dry the filter in an incubator at 40°C or under an infrared lamp. Place the filter in a scintillation vial with a toluene-based scintillant and measure ^{3}H radioactivity. Calculate the specific radioactivity of the RNA (cpm/µg).

13. Store the RNA solution frozen. For the class experiment, provide vials containing 3 mL solution of radioactive RNA at a concentration of 5 µg/mL. Any remaining RNA can be frozen at −20°C for subsequent use.

Growth of Plant Material for DNA Extraction

This process is exactly as described in this Appendix for Chapter 10.

Solutions and Chemicals

All the solutions and chemicals listed in this Appendix for Chapter 10, in the section entitled "Solutions and Chemicals," are required for DNA extraction and purification, *except* the DNA buffer, which has a different λ composition in this experiment. The following are also needed:

1. 100 mL Sodium tetraborate buffer, pH 9.2 (DNA buffer)

2. 10 L 2 × SSC (standard saline citrate)
   0.3M Sodium chloride
   0.03M Trisodium citrate, pH 7

3. Scintillation fluid Any toluene-based scintillation fluid will do, such as: 4 g 2,5-diphenyloxazole, 50 mg 1,4-bis(5-phenyl-2-oxazolyl)-benzene, dissolved in 1 L toluene

4. Actinomycin D stock solution. Dissolve at a concentration of 1 mg/mL in DNA buffer (described above) and store cold, protected from the light

5. 100 mL 5.0M Sodium hydroxide

6. Neutralizing medium. Make up the following reagents:

1 L 1.0M HCl
1 L 1M Tris-HCl, pH 7.8
2 L 3M NaCl

Mix the reagents in the following proportions immediately before use: HCl:Tris:NaCl (1:1:2) 4 L. Check that 20 mL of this solution gives a pH of 7.8 when mixed with 1 mL of 5.0M NaOH.

Apparatus and Equipment

This is as described in the Appendix for Chapter 10. In addition, the following items are needed:

Liquid scintillation counter
50 Scintillation vials and caps
Water baths at 37 and 70°C
Incubator at 40°C *or* an infrared lamp
Vacuum desiccator and vacuum pump
Vacuum oven to run at 80°C
One pair of forceps per pair of students
60 Nitrocellulose filters (13 mm diameter, 0.45 μm pore size)
50 Swinnex filter holders with 8 suitable test tube
racks for support
40 10- or 20-mL Disposable syringes

The experiment can be performed without a vacuum oven, but some of the DNA may wash off the filters during the hybridization reaction.

Chapter 12

General Preparation

Very little specialist or expensive equipment is required for this experiment and it is therefore suitable for quite large classes. The use of ^{32}P improves the data interpretation aspects of the experiment and gives the participants experience in using radioactivity. It is not essential to the experiment, however, and it may be omitted if circumstances dictate. Alternatively, [^3H]uridine (25–50 μCi/mL) can be used instead of ^{32}P, but this increases the cost of the experiment. If a radioisotope is being used, it is advisable to schedule the first day of the experiment for an afternoon, since a 4-h labeling period is required before the practical class starts.

Quite large amounts of radioactivity are used in this experiment, so the pea seedlings must be incubated with the radiolabel in a licensed laboratory. The bulk of the radioactivity is removed, however, before the seedlings are used in the experiment. Refer to the safety precautions pertaining to the use of radioisotopes described in Appendix I, "Safe Handling of Radioactivity."

Preparation of Plant Material

1. Five days before the experiment is scheduled, surface sterilize approximately 100 mL of dry pea seeds by soaking in sodium hypochlorite solution (1% available chlorine) and a drop of liquid deter-

gent for 15 min. Imbibe the seeds for 4 h in running water. Rinse the seeds several times in sterile
distilled water and germinate on autoclaved,
moist paper towels in a container with a lid in the
dark at 18–22°C.

2. Four hours prior to the beginning of the experiment, select approximately 50 seedlings with radicles approximately 1 cm long and place them,
radicle downward, in a sterile Petri dish. Add 10
mL ^{32}P-orthophosphate, 50 μCi/mL, in sterile distilled water containing 25 μg/mL streptomycin
and chloramphenicol. Incubate for 4 h at 18–
22°C.

3. Transfer the seedlings to a beaker and rinse in
several changes of distilled water to wash off the
bulk of the ^{32}P.

Day 1, Materials

1. Benchcote at 10 positions in the laboratory, or 10
spill trays.
2. Disposal receptacles for aqueous, organic solvent,
and solid radioactive wastes.
3. Disposable rubber gloves.
4. 20 Pairs of safety glasses.
5. 10 Mortars and pestles.
6. 50 Glass 15-mL centrifuge tubes and caps, with
racks.
7. Phenol mixture:

> 500 g phenol crystals
> 70 mL *m*-cresol
> 0.5 g 8-hydroxyquinolene
> 150 mL water

The phenol and *m*-cresol should be colorless; if
not they must be re-distilled. The solution is
intended to be water-saturated. Store in a dark
bottle at 4°C for up to 2 mo. The solution darkens in color with age owing to oxidation. Discard
the solution if the color darkens beyond light
brown. The *m*-cresol is an optional component

that acts as an antifreeze and additional deproteinizing agent. *Note: Solutions containing phenol are highly toxic; gloves and safety glasses should be worn.*

8. Detergent solution:

    1 g Sodium tri-isopropylnaphthalene sulfonate (TPNS)

    6 g Sodium 4-amino salicylate

    5 mL Phenol mixture

    Make to 100 mL in 50 mM Tris-HCl (pH, 8.5)

    Mix the TPNS with the phenol mixture before adding the other components. Store as for phenol mixture. Provide in an automatic dispenser or as an aliquot in a small bottle for each pair of students.

    Alternative detergent solutions can be used, such as the following, based on sodium dodecyl sulfate (SDS):

    0.15M Sodium acetate

    0.05M Tris-HCl (pH, 9.0)

    5 mM EDTA

    1% (w/v) SDS

    20 μg/mL Polyvinyl sulfate

9. The deproteinizing solution consists of phenol mixture and chloroform mixed 1:1 by volume, 200 mL. Provide in an automatic dispenser or as an aliquot in a small bottle for each pair of students. Store as you did the phenol mixture.

10. Absolute ethanol, 200 mL.

11. A Petri dish, scalpel, and pair of forceps for each pair of students

Day 2, Materials

1. Benchcote, disposal receptacles, and rubber gloves, as for day 1.

2. 10 Small chromatography columns. These can be easily made from Pasteur pipets plugged with glass wool and attached to a short piece of tubing. The flow rate can be regulated with a small screwclip on the tubing.

3. 200 Small test tubes to hold 1.5 mL.

4. 10 Test tube racks for above.

5. At least one UV spectrophotometer with cuvets.

6. 200 Scintillation vials and caps.

7. 2 Filter tower apparatus suitable for 2.5-cm filter disks.

8. 50 GF/C glass fiber disks, 2.5 cm.

9. 10 × 25 mg Oligo(dT)-cellulose swollen in 5 mL elution buffer.

10. Oligo(dT)-binding buffer, 400 mL:
    20 mM Tris-HCl (pH 7.5)
    0.5M NaCl
    1 mM EDTA
    0.2% w/v SDS

11. Oligo(dT)-elution buffer, 200 mL:
    20 mM Tris-HCl (pH 7.5)
    1 mM EDTA
    0.2% (w/v) SDS

12. 70% v/v Ethanol containing 0.5% w/v SDS, 200 mL.

13. Bovine serum albumin, 500 μg/mL, 10 mL.

14. 10% (w/v) Ice-cold trichloroacetic acid (TCA), 50 mL.

15. 5% (w/v) Ice-cold TCA, 500 mL.

16. A scintillation counter.

17. Sufficient scintillation fluid for 150 aqueous samples and 20 filters. If ^{32}P is used, water can be used as a scintillation fluid with the scintillation counter set for Cerenkov counting.

18. 20 Microcentrifuge tubes (1.5 or 0.5 mL) for storing nucleic acid aliquots prior to electrophoresis (optional).

19. Absolute ethanol, 200 mL

Chapter 13

General Preparation

Nucleic Acid Samples

If agarose gel electrophoresis is being used as an extension of the experiment described in Chapter 12, every student will have a pair of nucleic acid samples containing 3 and 5 μg. The lower loading gives sharper results for high mol wt RNA, but the higher loading is often necessary to observe low molecular weight RNA.

As an alternative, or in addition to these nucleic acid samples prepared from pea radicles, it is possible to run nucleic acid samples from other sources. Total nucleic acids can be prepared from the young shoots of pea seedlings grown in the light by the procedure described in Chapter 12. These samples will contain rRNA from the chloroplasts and will therefore produce more bands on electrophoresis. Alternatively, 16S and 23S RNA from *E. Coli* can be purchased from a molecular biology products supplier.

Solutions and Chemicals

Note: RNA is very susceptible to nuclease attack. All solutions that come into contact with the sample should be autoclaved whenever possible.

The solutions described below are sufficient for 10 gels. Depending on the total number of samples to be run, a greater or fewer number of gels can be cast and the volumes adjusted pro rata.

1. Two bottles of alcohol, for washing gel plates, in wash bottles. Industrial methylated spirits is quite satisfactory.

2. 1 L Gel-running buffer × 10-Tris $0.9M$, Na_2 EDTA 25 mM, boric acid $0.9M$, adjusted to pH 8.2 using HCl. This stock solution is 10 times its final concentration and should be autoclaved and stored at 4°C.

3. Sterile distilled water for diluting running buffer, 1 L/gel tank.

4. 10 × 100 mL Agarose solution—agarose (e.g., type 1, Sigma, or any equivalent preparation with a low coefficient of electroendosmosis), 1.1% (w/v) in single-strength, gel-running buffer. Autoclave to dissolve the agarose, then keep at 50°C in a stoppered vessel until needed.

5. 5 × 1 mL Gel-loading buffer: 50% (w/v) glycerol (or 20% sucrose), 0.4% (w/v) bromophenol blue in single strength gel running buffer, autoclaved.

6. 5 × 1 mL Ethidium bromide, 5 mg/mL in distilled water or single-strength gel-running buffer. *Wear rubber gloves when handling this compound; it is mutagenic.*

Equipment

Assume one of each item for every gel that is to be run, unless stated otherwise.

1. Glass casting plate, 23 × 12.7 cm, 3 mm thick, with its edges smoothed. Perspex plates are not suitable, since they tend to warp.

2. Zinc oxide tape, 12.5 mm wide.

3. Scissors.

4. Leveling table and spirit level (optional).

5. Electrophoresis tank, well-forming comb, and power pack, as described in this appendix for Chapter 3.

6. Adjustable, automatic micropipet, covering 5–30 μL.

7. Disposable tips for micropipet, sterilized by auto-claving (approximately 40 tips per tray, five trays per class of 20).

8. Microcentrifuge (at least 1/class).

9. Water bath, 50°C (at least 1/class).

10. Shallow tray or plastic box, for staining gel. This must be sufficiently large and shallow for easy access to the stained gel.

11. Pair of disposable gloves and safety glasses for each student.

12. Transilluminator and camera (optional) as described in this appendix for Chapter 3. For pro-tection against sunburn, the transilluminator should be screened by a perspex screen, 5 mm thick. *Safety glasses must also be worn to protect the eyes; ordinary spectacles are not sufficient.*

Denaturing Gels

If the RNA is to be denatured with glyoxal and dimethyl sulphoxide, the following modifications or addi-tions are required. The following materials are sufficient for 10 gels.

1. 1 L $0.1M$ NaH_2PO_4/Na_2HPO_4 (pH 7.0) phos-phate-running buffer $\times$ 10 (instead of Tris-borate running buffer).

2. 10 mL $0.03M$ Phosphate buffer made from the stock solution above.

3. 1 mL $6M$ Glyoxal (40% solution) deionized by passing through a mixed-bed resin (e.g., Bio-Rad Ag 501-X8) until its pH is neutral. Store at $-20°C$ in aliquots in tightly capped and completely filled tubes. Glyoxal readily oxides in air and this pro-cedure is essential before use.

4. 1 mL Dimethyl sulfoxide.

5. 10 mL Phosphate gel-loading buffer: 20% glycerol (or sucrose), $0.01M$ NaH_2PO_4/Na_2HPO_4 (pH 7.0) 0.4% bromophenol blue (instead of Tris-borate loading buffer).

6. One peristaltic pump/gel to circulate the running buffer during electrophoresis (optional).

7. Test tube rack for microcentrifuge tubes in a 50°C water bath.

Chapter 14

General Preparation

This section gives the quantities of all the materials required for a class of 20 students (working in pairs) to carry out the experiments. Detailed recipes are given for the various solutions required. It is important that the recipes for solutions are followed exactly and that the sources of chemicals are adhered to where possible.

Expt. 1 may not be possible for ten groups of two students. This experiment can either be performed using larger groups and dividing the wheat germ accordingly, as a demonstration, or prior to the practical class by the teacher. Similarly, groups of students can share polyacrylamide gels in Expt. 3.

It is important that all solutions and glassware used in these experiments are sterile. Solutions should be auto-claved (unless otherwise indicated) for 15 min at 15 psi. Glassware can be sterilized in an oven at 180°C for 3–4 h, preferably overnight. Furthermore, gloves should be worn by the students to reduce the possibility of contamination by ribonuclease. Guidelines for the handling and disposal of radioactive waste are given in Appendix I.

Expt. 1. Preparing the Wheat Germ Lysate

1. 200 mL Wheat germ grinding buffer.
    20 mM Hepes, pH 7.6
    100 mM Potassium acetate

333

      1 mM Magnesium acetate
      2 mM Calcium chloride (hydrated)
      6 mM Dithiotheitol (added after autoclaving)

2. 1 L Column buffer.

      10 mM Hepes, pH 7.6
      100 mM Potassium acetate
      4 mM Magnesium acetate
      1 mM Dithiothreitol (added after autoclaving)

3. Sephadex G25. Use Pharmacia, coarse, or medium grade (*not* fine or superfine).
Swell Sephadex by autoclaving in column buffer for 20 min at 15 psi.
Swell enough Sephadex to give 20 cm bed height (approximately 15.7 mL = 4.0 g Sephadex G25).
The column used should be 1 cm in diameter and 25 cm long, with a sintered glass support at the bottom end (a plug of glass wool will suffice).

4. 1 L Liquid nitrogen.

5. 100 g Wheat germ.

6. 10 Mortars and pestles, 4°C.

7. Sterile Pasteur pipets.

8. 20 × 15-mL Sterile glass centrifuge tubes (e.g., Corex tubes) to withstand 30,000g.

9. 10 Sterile spatulas.

10. 8 × 50-mL Rotor, 4°C

11. 10 Chromatography columns,

12. 10 10-mL Syringe barrels, or
1 box disposable pipet tips to hold 50 μL.

13. 10 Plastic universal bottles.

14. Cooled centrifuge, 4°C.

15. Freezer, −20°C (or −80°C) (or liquid N$_2$).

16. 10 Ice buckets and ice.

17. 10 Pairs of safety glasses.

Expt. 2. Optimizing the Magnesium Concentration for the Wheat Germ Lysate

1. Hepes, potassium acetate, 5 mL
Dissolve the Hepes in distilled water to 160 mM,

add potassium acetate to 720 mM, pH to 7.6 with KOH, sterilize by autoclaving.

2. ATP, creatine phosphate, 1 mL
   Dissolve ATP in sterile water and bring to pH 7.0 with KOH. (If pH is overshot, start again.) Add creatine phosphate.
   Final concentrations: ATP 40 mM (Sigma, A 7894)
   CP 320 mM (Sigma, P6502)
   Store at −20°C.

3. Spermidine, 1 mL (Sigma S 2501)
   Dissolve to 12 mM in sterile water. Store at −20°C.

4. Amino acid solution, 1 mL
   Use Sigma amino acid kit (LAA-21)
   Make a 20-mM stock solution in sterile water of each except tyrosine (10 mM). (Omit hydroxyproline and cystine, since these are amino acids obtained by posttranslational modification, and are thus not incorporated in in vitro synthesized proteins.) For the translation, use all the amino acids at 1 mM except the one that is added labeled. Store at −20°C.

5. Dithiothreitol, 1 mL (Sigma D 0632)
   Dissolve to 200 mM in sterile water. Store at −20°C.

6. Creatine phosphokinase, 1 mL (Sigma C 3755)
   Dissolve in 50% glycerol at 5 mg/mL. Store at −20°C.

7. GTP, 1 mL (Sigma G 5631)
   Dissolve to 10 mM in sterile water.
   Store at −20°C—short term.
   Store at −80°C—long term.

8. Radiolabeled amino acid.
   [^{35}S] Met, 9 μL   AMERSHAM SJ204
   or [^{3}H] Leu   AMERSHAM TRK 510
   or [^{3}H] mix of five amino acids:
   Pro, Leu, Phe, Lys, Tyr.   AMERSHAM TRK 550
   The second and third radiolabels are alternatives that may be used in laboratories not equipped to handle [^{35}S]. Alternative suppliers of radiochemicals may also be used, for example NEN.

9. 100 mM Magnesium acetate, 200 mL

10. Messenger RNA:
    Globin mRNA 35 μL (0.005 μg/mL) in sterile water, and/or TMV RNA 5 μL (0.1 μg/mL) in sterile water, and/or alternative mRNA (e.g., from the experiment described in Chapter 12).
    mRNAs are dissolved in sterile distilled water at the following approximate concentrations per translation:
    poly A$^+$ mRNA 0.001–5 μg
    The amount of mRNA used depends on its purity. If the mRNA is pure and contains one or a few species, then little is needed. Alternatively, if the mRNA is relatively impure or contains a large number of species, then a larger amount is needed.
    poly A$^-$ RNA 2–10 μg

11. Wheat germ extract from Expt. 1, 1 mL.

12. Sterile distilled water, 1 L.

13. 20% Trichloroacetic acid (TCA) plus 0.5% methionine, 100 mL.

14. Industrial methylated spirit (IMS):diethyl ether, 1:1, 2 L.

15. IMS, 2 L.

16. 5% TCA, 2 L.

17. 10 Micropipets, 20, 100, or 200 μL.

18. Sterile tips to fit above (20 racks of 200).

19. Sterile Eppendorf tubes (1.5 mL, approximately 300).

20. 10 150-mL beakers.

21. 10 Ice buckets, lids, and ice.

22. 10 21-Gage hypodermic needles.

23. 300 Filter paper rectangles (1 × 2 cm, approximately).

24. 10 Millipore forceps.

25. 10 Pins and polystyrene.

26. 1 Roll aluminum foil.

27. Paper tissues.

28. Paper towels.

29. 10 Glass rods.

30. 10 Pairs disposable latex gloves.

31. Waterbath 28°C and waterbath boiling (2 or more of each).
32. 10 Racks for waterbath to hold Eppendorf tubes.
33. 1 Or more microcentrifuge (to spin Eppendorf tubes).
34. 300 Scintillation vials and caps.
35. Solvent waste bottle.
36. Scintillation fluid and dispenser.
37. Scintillation counter.
38. Wash-up bowl, water, and Decon 90 detergent.
39. Fumehood and Benchkote.
40. Benchkote or radioactive spill trays at 10 places in the laboratory.
41. Disposable receptacles for radioactive waste.
42. Radiation safety monitor.

Expt. 3. Translation for Analysis of Products

The number of gels run will be determined by the number of translations carried out and the number of gel dryers available. Use the apparatus from Expt. 2, in addition to the following:

1. SDS polyacrylamide gels (*see* Chapter 15).
2. Plastic boxes to stain and destain gels.
3. Gel dryer.
4. Sheets of Whatman 3MM paper (or equivalent).
5. Clingfilm.
6. Glass plates (same size or larger than gel), 2/gel.
7. Masking tape.
8. Black polythene bag, 1/gel.
9. Freezer, −80°C (or less effective, −20°C).
10. X-ray film: Kodak X-O-MAT, XAR5. Developer: Kodak LX24. Fixer: Kodak FX40, diluted according to instructions.
11. Power packs.
12. 25 µL syringe.

13. Gel stain, 2 L.
    0.5% Kenacid blue
    0.1% Cupric acetate
    in 30% Industrial methylated spirit
        10% Acetic acid
        60% H_2O
    Stir to dissolve stain. Filter before use.

14. Gel Destain, 3 L.
    Use the items for gel stain (step 13), omitting
    Kenacid blue and cupric acetate.

15. "Cracking" buffer (for SDS PAGE gels)
        5 mL 1.25M Tris, pH 6.8
        2 g Sodium dodecyl sulphate
        40 g Sucrose
        10 mg Bromophenol blue
        1 mL β-mercaptaethanol
        1 mL 0.2M EDTA
    Make to 100 mL with distilled H_2O.

Chapter 15

General Preparation

SDS Gel Electrophoresis

The limiting factor in this experiment is the number of slab gel apparatus sets available to the students. Two students per apparatus set is convenient, but up to four students per apparatus set is manageable.

The following solution volumes are appropriate for one set of apparatus, and allowance has been made for duplication in case of error. However, since all the solutions (except ammonium persulfate) are stable for many months at 4°C, it may prove more convenient to prepare larger volumes and store them after use if the experiment is to be run more than once each year.

Note: Acrylamide is reported to be a neurotoxin. Therefore, take care when preparing acrylamide solutions. Polymerized acrylamide is NOT toxic. Unless otherwise stated, all solutions are stable for many months at 4°C.

1. Stock 50 mL acrylamide (30% acrylamide, 0.8% bis-acrylamide):
    Dissolve acrylamide monomer (75 g) and *N,N*-methylene bis acrylamide (2 g) in approximately 150 mL water. When dissolved, make the volume up to 250 mL, filter, and store at 4°C.

2. 20 mL 1.875*M* Tris-HCl, pH 8.8. Filter before storing.

339

3. 10 mL 0.6M Tris-HCl, pH 6.8. Filter before storing.

4. 5 mL 5% (w/v) Ammonium persulfate. (Make fresh as required, but solution will keep at room temperature for at least 1 wk.)

5. 5 mL 10% (w/v) SDS. Store at room temperature. In cold weather SDS can come out of solution but may be redissolved by standing the bottle in warm water.

6. Electrode buffer (0.05M Tris, 0.384M glycine, 0.1% SDS):

   Tris 12 g
   Glycine 57.6 g
   SDS 2 g

   Make to 2 L with distilled water. No pH adjustment.

7. ×2 Sample buffer:

0.6M Tris-HCl, pH 6.8	10.0 mL
SDS	1.0 g
Sucrose	10.0 g
Mercaptoethanol	0.5 mL
Bromophenol blue	5.0 mL
	of a stock
	0.5% solution

   Make to 50 mL and keep well stoppered.

8. Standard protein samples. Some or all of the proteins shown in Table 1 should be dissolved in ×1 sample buffer at 1 mg/mL and heated to 90–100°C for 5 min. They may be stored frozen indefinitely.

9. Unknown protein samples. Pure proteins should be dissolved at 1 mg/mL in ×1 sample buffer. Other useful "unknown" samples include:
   Milk: Add 50 μL of milk to 1 mL ×1 sample buffer and heat to 90–100°C for 5 min.
   Serum: Add 75 μL to 1 mL ×1 sample buffer and heat to 90–100°C for 5 min.
   Liver homogenate: Homogenize rat liver (≈15 g) in 40 mL of water for 1 min. Centrifuge in a bench centrifuge for 10 min. Add 250 μL of the

supernatant to 5 mL of ×1 sample buffer and heat to 90–100°C for 5 min.

All samples should be stored frozen once prepared.

10. Protein stain: 0.1% Coomassie Brilliant Blue R250 in 50% methanol, 10% glacial acetic acid (1 L). **Note:** Dissolve the stain in the methanol component first.

11. Protein destain: 10% methanol, 7% glacial acetic acid (2 L).

12. Gel apparatus. A number of gel electrophoresis apparatus are available commercially, but these can prove expensive if a large number of apparatus sets are required for class practicals. However, only modest workshop facilities are necessary for one to be able to construct a "homemade" perspex apparatus. At the time of this writing, an apparatus can be assembled for well under $100, the main cost being the platinum wire used as electrodes. A suitable design has been described by Studier (ref. 4 in Chapter 15), and full design details can be obtained from this reference. Plastic boxes are required for staining and destaining gels.

13. N, N, N', N'-tetramethylenediamine (TEMED)—single bottle for the total class, placed in a fume cupboard.

14. Microsyringes or automatic pipets (1/gel apparatus) capable of delivering 10–20 μL.

15. A power pack capable of delivering 50 mA at 400 V.

16. 100-mL Buchner flask and bung

17. Methylated spirits (IMS) in a wash bottle

18. A light box for viewing the gels

Silver Staining

Silver stain reagents should all be prepared the day they are required. The following solutions are required to stain a single gel.

100 mL 10% (v/v) Glutaraldehyde (SLR)
100 mL 5 μg/mL Dithiothreitol
100 mL 0.1% (w/v) Silver nitrate
100 mL Developer solution: 50 μL 37% formaldehyde in 100
    mL 3% sodium carbonate
10 mL 2.3M Citric acid

General Comments

Although this experiment gives the student experience in preparing and running acrylamide gels, the technique is being used here to determine the molecular weight of an "unknown" protein. This may of course be any protein of known molecular weight that is given to the student. Some suitable proteins and their molecular weight values are recorded in Table 1 in Chapter 15. However, it is also of interest to introduce a more complex unknown, such as milk, serum, or a liver homogenate. In this case, the student can comment on the complexity of the mixture and determine the molecular weight range of the components and the molecular weight of any major components.

If time is a problem with this practical class, it is possible to just prepare and load the gel, then run it at a lower current ($\approx$8 mA) overnight. However, if this method is used, it is unlikely that the students will be available next morning to dismantle the apparatus and stain the gel, so this task will have to be done by the technical staff. Whichever time scale is used, it is obvious that the students cannot record their results until 1 or 2 d after the practical class, although recording the results will only take a few minutes. However, the stained protein bands are quite stable and can be observed at any later date.

The silver stain should be carried out on a gel that has been previously stained with Coomassie Brilliant Blue, and exhaustively destained. It is therefore best for the students to silver stain a gel previously run by a different class, or to silver stain their own gels at a later date. It is of particular interest to note that appearance of many extra bands in areas of the gel that did not originally stain with Coomassie. The observed increase in sensitivity of this method over the Coomassie method is quite dramatic. For convenience it is useful if the first two steps of the silver stain procedure are carried out by the technical staff on the previous day.

Chapter 16

General Preparation

The number of gels to be prepared will depend on the number of electrophoresis tanks available. The following amounts are sufficient for preparing *one* IEF gel. Since up to 20 samples can be loaded onto one gel, it is reasonable for four students to share a gel. It is advisable that "spare" IEF gels are prepared, since gels are occasionally ruined when the glass plates are pried apart. Unused "spare" gels can be stored at 4°C for many weeks.

1. Glass plates, 22 × 12 cm. These should preferably be of 1 mm glass, but 2 mm glass will suffice.

2. PVC electric insulation tape. The thickness of this tape should be about 0.15 mm and can be checked with a micrometer. The tape we use is actually 0.135 mm.

3. Stock acrylamide solution (10 mL). Mix the following:

   3.88 g Acrylamide
   0.12 g Bis-acrylamide
   10 g Sucrose
   80 mL H$_2$O

   This solution may be prepared some days before the experiment and stored at 4°C.

4. Riboflavin solution (1 mL). This should be made fresh, as required. Stir 10 mg of riboflavin in 100 mL of water for about 20 min, then leave the

undissolved material to settle out (or briefly centrifuge).

5. Ampholyte solution (pH range 3.5–10.0) (0.5 mL). This is available commercially and can be stored for years at 4°C. However, take care not to introduce microbial contamination.

6. Twenty sample application papers: 0.5 × 0.5 cm squares of Whatman 1MM filter paper or similar. The square should be able to absorb 5–8 µL of solution.

7. Electrode wicks. These are 22 × 0.6 cm strips of Whatman No. 17 filter paper. However, any filter paper (or layers of) that provide a thickness of 2–3 mm will suffice.

8. Anolyte: $1.0M$ H_3PO_4 (10 mL).

9. Catholyte: $1.0M$ NaOH (10 mL).

10. Fixing solution: mix 150 mL methanol and 350 mL distilled water. Add 17.5 g sulfosalicylic acid and 57.5 g trichloracetic acid.

11. Protein stain (200 mL): 0.1% Coomassie Brilliant Blue R250 in 50% methanol, 10% glacial acetic acid. **Note**: Dissolve the stain in the methanol component first.

12. Protein destain (400 mL): 10% methanol, 7% glacial acetic acid.

13. A high-voltage power pack capable of delivering 1500 V at ≈10 mA is required with the IEF appartus.

14. Standard protein samples should be prepared as aqueous solution (1 mg/mL) and stored at −20°C. Should any proteins precipitate on freezing, they will need to be prepared fresh for each experiment.

15. A microsyringe capable of delivering 5–10 µL.

16. If students are collecting their own blood samples, a finger lance will be required, together with 70% alcohol for washing the skin prior to puncturing.

17. Plastic spillage tray

18. Access to a light box

Chapter 17

General Preparation

The following amounts are sufficient for *one* protein blotting experiment. Antibody solutions should be prepared just prior to being required.

1. Nitrocellulose paper, available commercially in large sheets.
2. Whatman No. 1 and 3MM filter paper.
3. Disposable diapers or paper towels.
4. Blotting buffer (100 mL):
    20 mM Tris, 150 mM glycine, 20% methanol, pH 8.3
    4.83 g Tris
    20.5 g Glycine
    400 mL Methanol
   Make to 2 L with water. Stable for weeks at 4°C.
5. Protein stain (20 mL):
    0.005% nigrosin in 2% acetic acid.
6. Destain (100 mL):
    2% acetic acid
7. 200 mL phosphate buffered saline (PBS):
    8 g NaCl
    0.2 g KCl
    1.15 g Na_2HPO_4
    0.20 g KH_2PO_4
   Make to 1 L with distilled water. Stable for weeks at 4°C.

345

8. A 1:200 dilution of rabbit anti-bovine serum albumin in PBS, containing 2% polyethylene glycol 6000 and 0.05% Nonidet NP-40 (20 mL).

9. A 1:1000 dilution of goat anti-rabbit IgG (conjugated to either horseradish peroxidase or alkaline phosphatase) in PBS containing 2% polyethylene glycol 6000 and 0.05% Nonidet NP-40 (10 mL).

10. 20 mL Of the appropriate enzyme substrate:
    A. *Horseradish peroxidase substrates*
    Two possible substrates are described below:
    1. Staining with 3-amino-9-ethylcarbazole.
    Prepare 20 mL of a 1% stock solution of 3-amino-9-ethylcarbazole in acetone (store at 4°C). *Note: 3-Amino-9-ethylcarbazole is a possible carcinogen—use with care.*
    The substrate solution is prepared by the student when required by mixing the following:

    > 0.8 mL Stock solution
    > 20 mL 0.05M Sodium acetate buffer, pH 5.0
    > 10 μL 30% Hydrogen peroxide

    2. Staining with 4-chloro-1-naphthol
    Prepare 20 mL of a stock 2% solution of 4-chloro-1-naphthol in methanol and store at 20°C.
    The substrate solution is prepared by the student when required by mixing the following:

    > 0.6 mL Stock solution
    > 20 mL 0.05M Tris-HCl, pH 7.4
    > 10 μL 30% Hydrogen peroxide

    Substrate A gives a reddish-brown insoluble dye deposit in the presence of horseradish peroxide. Substrate B gives a blue deposit.
    B. *Alkaline phosphatase substrate*
    5-Bromo-4-chloro-3 indolylphosphate (2 mg/mL) in 1M diethanolamine, pH 9.8 (20 mL). This substrate gives a blue product. Other commonly used alkaline phosphatase substrates give a yellow product that is not as easy to detect on a nitrocellulose sheet.

11. 100 mL 2% Tween in PBS

12. Small dishes for staining and destaining

Chapter 18

General Preparation

1. Sephadex G-200 SF should be preswollen in 0.9% NaCl for at least 3 d prior to the experiment. A swollen volume of 250 mL will be sufficient fo ten plates (10 × 20 cm). Any unused gel can be stored in saline at 4°C, but should have sodium azide added to give a final concentration of 0.1% to act as a preservative.

2. Glass plates (10 × 20 cm) should be cleaned with detergent and rinsed before use.

3. Standard protein samples (*see* Table 1 in Chapter 18) and an "unknown" protein should be prepared at 15 mg/mL in 0.9% sodium chloride.

4. Eluent: 0.9% sodium chloride (100 mL/apparatus).

5. Protein stain: 0.2% Ponceau S in 10% trichloroacetic acid (100 mL/apparatus).

6. Destain: 2% acetic acid (200 mL/apparatus).

7. 3MM filter paper.

8. Whatman No. 1 filter paper for blotting the gel.

Chapter 19

General Preparation

Students should work individually in this experiment. However, since such small volumes are required, many of the reagents are best centralized rather than being divided between the students. Unless specified, the following are for the *total* class of 20 students.

1. Ground-glass stoppered test tubes (approximately 65 × 10 mm, e.g., Quickfit MF 24/0). All reactions are carried out in this "sequencing" tube. The sample to be sequenced (≈100 μg of peptide) should be dried down from an aqueous solution in the bottom of this tube; one tube (plus sample) for each student.

   Many simple peptides (tripeptides, pentapeptides, and so on) are commercially available, and are quite inexpensive. Avoid samples in which either of the first two residues are histidine, arginine, cysteine, or tryptophan; these residues are somewhat more difficult to detect and should be avoided. It may also be easier to avoid the complications that arise when there are adjacent hydrophobic residues (see Chapter 19).

2. 50% Pyridine (10 mL) (aqueous, made with A.R. pyridine). Store under nitrogen at 4°C, in the dark. Some discoloration will occur with time, but this will not affect results.

3. Phenylisothiocyanate (5% v/v) in pyridine (A.R.) (10 mL). Store under nitrogen at 4°C in the dark. Some discoloration will occur with time, but this will not affect results. The phenylisothiocyanate should be of high purity and is best purchased as "sequenator grade." Make up fresh about once a month.

4. Water saturated n-butyl acetate (100 mL—divide into five lots and distribute between groups of students). Store at room temperature; stable indefinitely.

5. Anhydrous trifluoroacetic acid (TFA) (10 mL). Store at room temperature, under nitrogen, in a fume cupboard.

6. Dansyl chloride solution (2.5 mg/mL in acetone) (5 mL). Store at 4°C in the dark. This sample is stable for many months if water vapor is eliminated. The solution should be prepared from concentrated dansyl chloride solutions (in acetone) that are commercially available. Dansyl chloride available as a solid invariably contains some hydrolyzed material (dansyl hydroxide), which makes the preparation of accurate solutions difficult.

7. Sodium bicarbonate solution (0.2M, aqueous) (5 mL). Store at 4°C. Stable indefinitely, but check periodically for signs of microbial growth.

8. 5N HCl (aqueous) (5 mL).

9. Test tubes (50 × 6 mm) referred to as "dansyl tubes"; two/student.

10. Polyamide thin-layer plates (7.5 × 7.5 cm); two/student. These plates are coated on both sides, and referred to as "dansyl plates." Each plate should be numbered with a pencil in the top corner of the plate. The origin for loading should be marked with a pencil 1 cm in from each edge in the lower left-hand corner of the numbered side of the plate. The origin for loading on the reverse side of the plate should be immediately behind the loading position for the front of the plate, i.e., 1 cm in from each edge in the lower right-hand corner.

11. Chromatography tanks capably of holding the polyamide plate (7.5 × 7.5 cm) are required. A minimum of three tanks, containing the following solvents, are required:

Solvent 1: Formic acid:water, 1.5:100, v/v
Solvent 2: Benzene:acetic acid, 9:1, v/v
Solvent 3: Ethyl acetate:methanol:acetic acid, 2:1:1, v/v/v

12. An acetone solution containing the following standard dansyl amino acids: Pro, Leu, Phe, Thr, Glu, Arg (each approximately 50 μg/mL). This solution is stable for many years at $-20°$C. Each student uses 2 μL at each experiment. Dansyl amino acids are available commercially, but may also be prepared from the free amino acids by scaling up the dansylation procedure (steps 11–15) described in the protocol section. (The acid hydrolysis step, although not absolutely necessary, is included to convert any unreacted dansyl chloride to dansyl hydroxide.)

13. One or more UV sources, either long wave (365 μ) or short wave (254 μ). Short wave is preferred.

14. A vacuum desiccator (containing NaOH pellets and a beaker of P_2O_5), capable of being partially immersed in a 60°C waterbath, and a suitable vacuum source. One such setup is suitable for the whole class, but two or more would reduce congestion.

15. Vacuum pumps used for this work should be protected by cold traps. Considerable quantities of volatile organic compounds and acids will be drawn into the pump if suitable precautions are not taken. A water pump can be used in place of a mechanical pump, but drying times are generally longer, depending on the efficiency of the pump.

16. Microsyringes and autodispensers capable of delivering 200 μL, and microsyringes capable of delivering 1.0 μL.

17. A nitrogen supply

18. A 50°C waterbath

19. 37 And 105°C ovens

Chapter 20

General Preparation

The main costs of this experiment are for the reagents, the sera, and specific antisera. It may, therefore, be impracticable to supply each pair of students with a set of reagents. It is possible, however, to supply one set of reagents for all students, with a fully labeled hypodermic syringe alongside each reagent.

It should be mentioned that the handling of any human serum products (as in the following Expts. 3 and 4) presents some hazard, e.g., hepatitis or AIDS. Thus, unless the source of serum can be guaranteed free of such hazards, it might be safer to replace the human serum samples with another serum, e.g., guinea pig.

The following reagents are sufficient for ten pairs of students doing all four experiments:

Phosphate Buffered Saline (PBS) 0.01M, pH 7.2

Make up to 1 L with deionized water.

$Na_2HPO_4 \cdot H_2O$	0.35 g
$NaH_2PO_4 \cdot 12H_2O$	2.68 g
NaCl	8.52 g

Preparation of the Agarose Gel for Day 1

Ten pairs of students will require a minimum of 250 mL molten agarose (allowing a little for accidents) for the four experiments.

Prepare 1% agarose in $0.01M$ phosphate buffered saline, pH 7.2 (PBS), by dissolving 2.5 g of agarose in 250 mL PBS and (a) heating in a boiling water bath, or (b) autoclaving for 5–10 min at 15 psi. When molten, transfer to a 50°C water bath. For ease of student access, it might be useful to divide the agarose into two or more aliquots and place them in separate water baths.

Note: It is possible to prepare the agarose gel in advance and melt by steaming for a short period (5 min at 15 psi) 1 h before the commencement of the experiment.

Materials for Day 1

1. 250 mL 1% Agarose in PBS (0.01M) at pH 7.2
2. Two or more waterbaths at 50°C
3. 40 Clean microscope slides
4. 2 Leveling tables (optional)
5. Templates for well pattern (may be taken directly from diagrams in the protocol)
6. 20 Sterile pasteur pipets (230 mm) and sterile 5- and 10-mL pipets
7. 20 Rubber teats
8. 14 Hypodermic syringes (1.0 mL) fitted with 25G needles (16 mm), labeled—one for each reagent (specific antisera and sera)
9. Whatman No. 1 filter paper
10. 4 Sandwich boxes with lids (to hold altogether a total of 40 slides)
11. Reagents for Expts. 1, 2, 3, and 4 (obtained from Sigma Chemical Co. or Miles Laboratories, Research Products Division) All dilutions are made in PBS:

Expt. 1

20 mL Bovine serum albumin at 5.0 mg/mL
20 mL Anti-bovine serum albumin
20 mL Human serum albumin, 5.0 mg/mL
20 mL Human serum albumin, 5.0 mg/mL

Expt. 2

20 mL Calf serum at 1/10
2.0 mL Anti-bovine serum albumin (undiluted)
2.0 mL Anti-bovine γ-globulin (undiluted)

Expt. 3

> Calf serum at 1/10, 1/100, and 1/1000 (20 mL each)
> Horse, rabbit, and human serum at 1/10 (20 mL each)
> 2.0 mL Anti-calf serum (undiluted)

Expt. 4

> 2.0 mL Anti-calf serum (undiluted)
> 2.0 mL Anti-horse serum (undiluted)
> 2.0 mL Anti-human serum (undiluted)
> 2.0 mL Unknown 'serum *x*' at 1/10 ('*x*' must obviously be either calf, human, or horse serum)

It must be noted that Expts. 1, 2, 3, and 4 can be performed individually. The reagent volumes, although given for each individual experiment, are sufficient for all four experiments. A box with oblique lighting can be used to view the unstained slides.

Materials for Day 2

1. 10 Sandwich boxes or plastic boxes for staining/ destaining
2. 10 × 100 mL Phosphate buffered saline (0.01*M*)
3. 10 × 100 mL Staining solution:
   > 0.2 g Coomassie Brilliant blue
   > 20 mL Glacial acetic acid
   > 100 mL Methanol
   > 80 mL Distilled water

   Note: It is advisable to first dissolve the stain in the methanol, and then add the acetic acid and distilled water.
4. 10 × 200 mL Destaining solution:
   > 20 mL Glacial acetic acid
   > 100 mL Methanol
   > 80 mL Distilled water
5. A light box

Chapter 21

General Preparation

Because of the small volumes of reagents involved, it is easiest to centralize most reagents, rather than divide them among students. For 20 students producing ten rocket and ten 2-D gels, unless stated otherwise, the amounts given below are required for the total class.

The amount of antiserum used in the gel depends of course on the antibody titer, and should be determined by trial and error when first setting up this experiment. The amounts quoted in the "Methods" section in Chapter 21 are the volumes of commercially available antisera (Miles) used in our experiments. These values should not vary much for antisera from different sources; once the level has been determined for one source, it is not necessary to test later batches from the same source.

1. 750 mL 0.06M Barbitone buffer, pH 8.4, for *each* electrophoresis apparatus (diluted 1:1 with distilled water for use in electrophoresis tanks).

    10.3 g Sodium barbitone
    1.84 g Barbitone

    Make to 1 L with distilled water.

2. Bovine serum (0.5 mL), undiluted, for 20 immuno-electrophoresis. (not necessary if students are analyzing their own serum.) Commercially available. Store frozen at −20°C in 1-mL aliquots.

3. Bovine serum containing bromophenol blue. Prepare by adding 50 μL 1% aqueous bromophenol blue to 1 mL of serum. Store frozen in 1-mL aliquots.

4. Antisera:

 (a) Rabbit anti-bovine serum
 (b) Rabbit anti-whole human serum
 (c) Rabbit anti-bovine serum albumin

These are all available commercially and should be reconstituted and stored according to the supplier's instructions.

5. Glass plates 5 × 5 cm. Preferably, the glass should be as thin as possible, but 2-mm window glass is suitable.

6. 2% Aqueous agarose. Prepare prior to the practical class by boiling, then place in a 52°C waterbath, where it can remain until the practical class starts. This solution is rather viscous and is best dispensed using a pipet with the point cut back to provide a larger orifice.

7. The following dilutions, in $0.03M$ barbitone buffer, of a stock (4% aqueous) solution of bovine serum albumin.

  1:1000
  1:750
  1:500
  1:400
  1:300
  1:200
  1:100
  1:75

These dilutions should be divided into 1-mL aliquots and stored frozen.

8. A 1:400 dilution of bovine serum as the test solution. Again this should be stored frozen in 1-mL aliquots.

9. Each pair of students should have a microsyringe capable of delivering 1 μL of sample.

10. Strips of Whatman No. 1 filter paper, 5 cm wide.

11. A flat bed electrophoresis apparatus with cooling plate. Most commercially available apparatus will hold four 5 × 5 cm plates.

12. Protein stain: 0.1% Coomassie Brilliant Blue R250 in 50% methanol. 10% glacial acetic acid (1 L).
    Note: Dissolve the stain in the methanol component first.

13. Protein destain 10% methanol, 7% glacial acetic acid (2 L).

14. If students provide their own serum samples, each donor will need a finger lance and a small microfuge tube (≈0.5 mL). A microfuge is also required.

Chapter 22

General Preparation

The Organism

The organism used in these experiments is *E. coli* strain CSH62, obtainable from any National Culture Collection. It grows slowly on glycerol and readily induces β-galactosidase when lactose is presented as an alternative carbon source (to glycerol).

Growth Medium

The *E. coli* is grown in a minimal medium containing 10.5 g dipotassium hydrogen orthophosphate, 4.5 g potassium dihydrogen orthophosphate, 1.0 g ammonium sulfate, 0.5 g sodium citrate, 0.21 g magnesium sulfate (hydrated), 0.1 cm^3 1% vitamin B$_{12}$, and 5 g glycerol/L of distilled water. The medium is dispensed (50 cm^3 aliquots) into 250-cm^3 cotton wool stoppered Erlenmeyer flasks and autoclaved to sterilize it prior to use.

Growth of Overnight Cultures

From a stock slope of freeze-dried culture of *E. coli* GSH62, inoculate a flask of culture medium prepared as above early in the day prior to the experiment being carried out. Incubate in a shaking incubator at 37°C and after 6–8 h, aseptically subculture from it 10 more flasks of medium for class use. Incubate overnight in the shaking incubator at 37°C.

Equipment Required

The group as a whole (20 students working in pairs) will require:

1. 10 Stop clocks
2. 10 Ice baths
3. 10 Vortex mixers
4. Waterbaths at 37°C
5. Shaker-incubator space (60 spaces in total)
6. 10 Bunsen burners
7. 60 Sterile 1.0 cm^3 pipets
8. 1400 Test tubes (140 per pair)
9. 10 Cylinders containing bleach or other sterilant for used pipets
10. Disposal arrangements for spent cultures
11. Spectrophotometers and glass cuvettes
12. 10 Light microscope and hemocytometer (or other cell counting chamber) or an electronic cell counter

Reagents Required

The group as a whole (20 students working in pairs), allowing for a small excess, will require:

1. 6 Overnight cultures of E. coli, grown as in the section "Growth of Overnight Cultures," above
2. 60 250 mL Erlenmeyer flasks, each containing 50 cm^3 minimal medium supplemented with 0.5% glycerol (see the section "Growth Medium" for details of the medium)
3. 100 cm^3 0.1% Chloramphenicol (0.1 g in 100 cm^3)
4. 50 cm^3 Toluene
5. 150 cm^3 0.1% Orthonitrophenol β-galactoside (ONPG) in 0.05M phosphate buffer pH 6.8 (1.5 g in 1500 cm^3)
6. 150 cm^3 0.01M Lactose (0.54 g in 150 cm^3)
7. 150 cm^3 0.01M Isopropylthiogalactoside (IPTG) (0.36 g in 150 cm^3)
8. 1000 cm^3 1.0M Sodium carbonate (106 g in 1000 cm^3)
9. 250 cm^3 0.0001M Orthonitrophenol (ONP) (0.0035 g in 250 cm^3)
10. 250 cm^3 0.05M Phosphate buffer pH 6.8

Chapter 23

General Preparation

The Organism

The organism used in these experiments is *Chlorella fusca* CCAP 211/8p obtained from the Culture Collection for Algae and Protozoa, Storey Way, Cambridge, UK, or from the author at nominal cost.

Maintenance of Stocks of the Organism

The organism is maintained in a liquid growth medium containing 7.30 g potassium dihydrogen orthophosphate, 2.32 g dipotassium hydrogen orthophosphate, 0.40 g magnesium sulfate (hydrated), 0.75 g ammonium nitrate, 0.40 g trisodium citrate, 0.002 g ferrous sulfate hydrated, and 1.0 mL Arnons A_4 micronutrient solution in 1 L glass distilled water. [Arnons A_4 contains in 250 mL distilled water, 2.864 g boric acid, 1.810 g manganese chloride (hydrated), 0.222 g zinc sulfate (hydrated), and 0.008 g copper sulfate (hydrated)]. Aliquots (15 mL) are dispensed into universal bottles, sterilized by autoclaving, and kept in a cold room prior to use.

Subcultures should be prepared every 8–10 wk by adding approximately 1 mL culture to 20 mL fresh medium using aseptic conditions. The caps are lightly replaced and the cultures incubated at room temperature in the light, preferably adjacent to a fluorescent tube.

Growth of Cultures for Experimental Use

There are several ways in which cultures for experimental use can be prepared. This will depend on facilities available. The easiest method, which requires the simplest equipment and minimum preparation, is given, with notes in parenthesis indicating alternatives.

Prepare 500 mL of culture medium, as in the section "Maintenance of Stocks of the Organism," above, in a 2-L Erlenmeyer flask. Autoclave, cool, and add the contents of a stock culture. Fit the flask with a sterile rubber bung through which is fitted glass tubing for aeration (Dreschel bottles may be used if available). Place the flask in an illuminated water bath (or an illuminated growth room) at 25°C. A suitable light intensity is 6000 lx, but this can be varied considerably. Aerate the cells with a mixture of 2% carbon dioxide in air (if this is not available, use 5% carbon dioxide in air, which is more readily available. If neither is available, the cells may be aerated vigorously with air, though growth will take longer.) The aeration gas should be warmed, humidified, and filtered by passing through a bottle of sterile distilled water and a sterile cotton wool filter prior to passing through the culture.

The cells are grown until approaching the end of the logarithmic phase of growth. The time taken to reach this stage will depend upon the growth conditions and can be determined by asceptically removing 2.0-mL samples at 24-h intervals, determining absorbance at 665 nm and plotting a graph of absorbance against time. Clearly, it will be necessary to carry out a pilot experiment to decide how long the cultures will need to be grown before this stage is reached under the growth conditions employed.

Once the end of the logarithmic phase is reached, aseptically add 50 mL of culture to each of eight sterile 250-mL Erlenmeyer flasks labeled A–H. Incubate each culture as follows:

A. light $+ CO_2$
B. light $- CO_2$
C. light $+ CO_2 +$ acetate
D. light $- CO_2 +$ acetate
E. dark $+ CO_2$
F. dark $+ CO_2 +$ acetate
G. dark $+ CO_2 +$ glucose
H. dark $+ CO_2 +$ acetate $+$ glucose

For cultures grown in the presence of carbon dioxide, the same conditions can be used as for the growth of the bulk culture (it will make no difference if air or CO_2-enriched air is used, since air contains enough CO_2 to maintain autotrophic growth). If growth conditions require CO_2-free air, this may be achieved by passing non-CO_2 enriched air through a bottle of 10% sodium hydroxide prior to its passage through the culture. For acetate-containing cultures, add a solution of sterile sodium acetate (pH, 6.5) to give a final concentration of 0.2% w/v (i.e., for 50-mL culture, add 0.1 g in 1 mL). For glucose-containing cultures, G and H, also add glucose to give a final concentration of 0.2% w/v. For dark conditions, wrap the flasks in aluminum foil to exclude light. Incubate all the cultures for 24 h in these conditions.

Harvesting Cultures and Preparation for Enzyme Assay

After incubation, the cells are harvested by centrifugation at 500g for 5 min and washed twice with 0.067M phosphate buffer, pH 7.5, and finally resuspended in 50 mL buffer. C. fusca cells are noted for the difficulty in disrupting them to release enzymes. However, the enzyme isocitrate lyase can readily be assayed in whole cells that have been subjected to freezing for a period of at least 24 h. Thus, after harvesting, the cells are placed in a deep freeze at between −5 and 15°C. They may be kept frozen for several weeks without noticeable deterioration of enzyme activity. This can be useful since larger "batches" of cultures could be grown if the experiment will be repeated several times with different groups of students. However, once thawed for use the enzyme preparation cannot be successfully refrozen and kept.

Equipment Required

The group as a whole (20 students working in pairs) will require:

10 Stop clocks
10 Ice baths
Waterbaths at 30°C
Spectrophotometers and cuvets (one per pair if possible, or as many as is available)

An oven set at 105°C
Desiccators
600 Test tubes (60 per pair)
160 Centrifuge tubes (16 per pair)

Reagents Required

The group as a whole (20 students working in pairs and allowing for some excess) will require:

0.067M Phosphate butter, pH 7.5 (500 mL)
0.15M Magnesium sulfate (hydrated) (1.849 g in 50 mL)
0.24M Trisodium isocitrate (1.764 g in 25 mL)
0.06M Reduced glutathione (0.9220 g in 50 mL)
15% Trichloroacetic acid (18.75 g in 125 mL)
0.1% Dinitrophenylhydrazine in 2M hydrochloric acid (0.25 g in 250 mL)
2.5M Sodium hydroxide (100 g in 1 L)
0.004M Sodium glyoxylate (standard solution) (0.192 g in 50 mL)

All these reagents can be prepared in advance and stored in a cold room until required.

Chapter 24

General Preparation

Requirements are given for ten pairs of students.

Strains

For purposes of illustration, this exercise has been presented using strains BSC 465/2a and FM 811c, which we originally obtained from S. Oliver. However, it must be emphasized that this method is a general approach to fusing otherwise incompatible yeast strains, and thus a wide variety of like mating-type pairs of strains could be used. In fact, with a class it is probably desirable to have different groups working with various combinations of strains. The only limitation is that some strains are either difficult to protoplast or regenerate, but this can be easily tested when planning the practical class.

The strains themselves are simply prepared by growing in shaking culture at 30°C overnight. The YEPD medium (2% yeast extract, 1% peptone, 2% glucose) is suitable for this purpose. Each pair of students will require 10 mL of each culture.

Solutions

Quantities given are for ten pairs of students. **Sterilize by autoclaving.**

363

1. 250 mL Sterile distilled water
2. 100 mL Reducing buffer
   1.2M Sorbitol, 25mM
   EDTA, 50 mM dithiothreitol, pH 8

4. 60 mL Enzyme buffer
   1.2M Sorbitol, 0.01M
   EDTA, 0.1M sodium citrate, pH 5.8
5. 20 mL Enzyme preparation
   20 mg/mL Novozym 234 (Novo, Copenhagen,
   Denmark) in sterile distilled water. Do not autoclave. As an alternative, β-glucuronidase (Sigma)
   or zymolyase (Kirin Brewery, Tokyo, Japan) can
   be used.
6. 10 mL 1% w/v SDS in water
7. 600 mL Sorbitol–CaCl$_2$ solution
   1.2M Sorbitol, 10 mM CaCl$_2$
8. 20 mL PEG–CaCl$_2$ solution
   20% polyethylene glycol 4000, 10 mM CaCl$_2$, 1
   mM Tris, pH 7.5

Note that different batches of PEG give differing fusion
efficiencies. It may be worthwhile to experiment with the
percentage used in this solution. In some procedures,
figures ranging between 15 and 30% are used.

Media

1.5 L Minimal agar
0.67% Difco yeast nitrogen base, 2% glucose, 1.2M Sorbitol, 2% agar. Sterilize by autoclaving and keep molten at
45°C. An alternative, cheaper, minimal media is that of
Wickerman (1).

1 mL of each of the following supplements:

20 mg/mL, filter sterilized. In this exercise adenine, histidine, uracil, and lysine are required, but this will vary
with strains.

Equipment

No unusual equipment is required, although it is desirable for centrifugation to be performed in a refrigerated centrifuge.

1. 1 (or more) Bench centrifuge
2. 20 Sterile, disposable plastic centrifuge tubes
3. 100 Small glass tubes
4. 200 Sterile glass universals
5. 10 Microscopes
6. 2 × 30°C Waterbaths
7. 4 × 45°C Waterbaths
8. 30°C Incubator
9. Sterile 10- and 1-mL pipets
10. 10 × Ice buckets

Yeast Transformation

The present exercise can be easily modified to include a demonstration of yeast transformation. In this case the method closely follows that of Beggs (2).

Strains

Yeast strains offering good transformation frequencies are somewhat rarer than those offering good protoplast fusion results. We have had success using the strain A3617C (genotype a *his3 gal2*) and the plasmid YEp6 (Amp^r, *his3*$^+$) in demonstration experiments. Both the strain and the plasmid originate from the laboratory of R. Davis. In this case, it is the transfer of the his$^+$ gene and subsequent ability to grow on unsupplemented minimal media that is selected for.

Protocol Amendments

Preparation of the protoplasts is unchanged. The washed protoplasts are finally resuspended in 0.1 mL sorbitol–$CaCl_2$, about 1 μg of plasmid DNA added (in 20 μL of its storage buffer), and left at room temperature for 15 min before adding the 1 mL of PEG–$CaCl_2$.

The only other change is a "revival" step, after fusion and before plating. At the end of the fusion period the cells are sedimented and then resuspended in 0.1 mL sorbitol–$CaCl_2$, plus 0.05 mL YEPD media including 1.2M sorbitol. This mixture is incubated for 20 min at 30°C. The volume is then made up to 2 mL with 1.2M sorbitol, and it is plated in supplemented and unsupplemented minimal media as already described.

Result

After 4–5 d incubation, supplemented media should reveal 1–20% regeneration. In general, up to one in 50 of these will be shown to be transformants by growth on the unsupplemented plates

References

1. Wickerham, L.J. (1946) A critical evaluation of the nitrogen assimilation tests commonly used in the classification of yeast. *J. Bacteriol.* **52**, 293–301.
2. Beggs, J.D. (1978) Transformation of yeast by a replicating hybrid plasmid. *Nature* **265**, 106–108.

Chapter 25

General Prepartion

Cultures

The three strains used in this exercise are available directly from the Ames Laboratory (Biochemistry Department, University of California, Berkeley, CA 94720, USA). They will also supply reprints of ref. 2 in Chapter 25, which gives details of the procedures for testing and storage of these strains.

Before the practical class, the cultures are grown overnight in shaken liquid culture at 37°C. Oxoid nutrient broth No. 2 is the recommended medium for this purpose. Each pair of students will require under 1 mL of culture; thus no more than 10 mL need to be grown up for a class of 20. If the practical class is in the afternoon, it is advisable to remove the cultures from the incubator first thing in the morning and store them on ice. If this is done, allow them to warm up to room temperature before use.

Media

Minimal Agar

The basic medium used is Vogel-Bonner minimal. Each pair of students will require 24 plates. Since each plate is recommended to have at least 25 mL of medium, for a class of 20 students at least 6 L of medium is required. The medium is made up as follows:

367

200 mL Salts solution ($\times 50$)
140 mL Warm distilled water
2 g Magnesium sulfate ($MgSO_4$, 7 H_2O)
20 g Citric acid monohydrate
100 g Potassium phosphate (K_2HPO_4)
35 g Sodium ammonium phosphate
  ($NaHNH_4PO_4$, $4H_2O$)

Salts are added and dissolved completely in 'the order shown using a magnetic stirrer. The volume is then made up to 200 mL and the solution autoclaved.

1 L Vogel-Bonner minimal agar
15 g Agar
930 mL Distilled water
20 mL $50\times$ Salts solution
50 mL 40% glucose

The agar is added to the distilled water and autoclaved. The salt solution and sterile glucose are added after autoclaving, the solutions mixed, and the plates poured.

Overlay Agar

Autoclave 0.6% agar and 0.5% NaCl in 50 mL aliquots. Present to the student in a 45°C waterbath. Each pair requires one 50-mL bottle.

Preparation of S-9

Induction of Rats

The recommended procedure is to use Arocolor 1254 (Analabs, North Haven, Connecticut, USA). This is diluted in corn oil to a concentration of 200 mg/mL and a single intraperitoneal injection of 500 mg/kg is administered to a 200–250-g male rat 5 d before sacrifice. Full details of this procedure are found in ref. 2 in Chapter 25.

For class purposes it is generally satisfactory to use rats induced with sodium phenobarbital. This is simply added at 0.1% to the drinking water for 1 wk before sacrifice. Although this treatment is regarded as producing a less satisfactory activation, it is generally more convenient. *Arocolor is itself carcinogenic and hence presents handling difficulties.*

If no animal handling facilities are available, S-9 can be obtained commercially from a number of sources, again listed in ref. 2 in Chapter 25.

Preparation of Rat Liver S-9 Fraction

All steps of this procedure are carried out at 4°C using cold, sterile solutions and glassware. The rat is killed by a blow to the head and the liver quickly removed and placed in a preweighed beaker containing 15 mL 0.15M KCl. After weighing, the liver is washed several times in fresh, cold KCl. The liver is transferred to a beaker containing 3 mL/g liver weight of cold 0.15M KCl and the liver minced with sterile scissors. The tissue is then homogenized in a Potter-Elvehjem apparatus with a Teflon pestle. The homogenate is then centrifuged for 10 min at 9000g, and the supernatant collected. This can be distributed into 2 mL aliquots in small tubes, quickly frozen, and stored at −80°C until needed.

Preparation of S-9 Mix

The S-9 fraction is thawed out immediately prior to use. Stock solutions of NADP (0.1M) and glucose-6-phosphate (1M) are prepared with sterile water in sterile tubes. These are stored frozen. Stock salt solution (0.4M $MgCl_2$, 1.65M KCl) and phosphate buffer (0.2M, pH 7.4) are autoclaved and kept in the refrigerator. The S-9 mix is made fresh (50 mL) each day as follows:

25.0 mL 0.2M Phosphate buffer
2.0 mL S9 fraction
1.0 mL Salt solution
).25 mL 1M Glucose-6-phosphate
2.0 mL 0.1M NADP
19.75 mL Sterile distilled water

Each pair of students will require 4.5 mL, hence 50 mL should be adequate for 20 students. The S-9 mix is kept on ice until required.

Preparation of Additives and Mutagens

50 mL Histidine–biotin supplement
0.5 mM histidine, 0.5 mM biotin, sterilize by autoclaving.

Mutagens (only minute quantities are required)
    Methyl methane sulfonate (used in original liquid form)
    9-Aminoacridine (1 mg/mL dissolved in ethanol)
    2-Aminoanthracene (1 mg/mL dissolved in DMSO)
Test compound (1 mg/mL should be a convenient concentration for
    spotting 10 μg on a plate)

Equipment

    45°C Waterbaths (Convenient access for each pair of students
        is essential. The overlay agar and 24 sterile universals per
        pair are set up in such baths.)
    37°C Incubator
    0–20 μL Micropipet and sterile tips (for each pair of stu-
        dents)
    Sterile pipets (1 mL and 10 mL)
    Pipet filler (one per pair)
    Disposal containers for contaminated pipets and glassware
    Disposal beaker for micropipet tips

INDEX

371